WORLD WILDLIFE HABITATS

Marshall Cavendish
New York Toronto Sydney

VOLUME 2

Published by Marshall Cavendish Corporation
2415 Jerusalem Avenue
North Bellmore
NY 11710 USA

Editorial Consultant: Roger Few
Co-ordinating Editor: Victoria Egan
Sub Editors: Milk Brown
Judith Devons
Paula Field
Peter Lowis
Designers: Peter Byrne
Sarah Hooker
Editorial Manager: Reg Cox

Reference Edition 1992

Typeset in Hong Kong by Best-set Typesetter Ltd
Printed and bound in Italy by LEGO Spa Vicenza

Library of Congress Cataloging-in-Publication data

World wildlife habitats. — Reference ed.

p. cm.
Companion to the Marshall Cavendish international wildlife encyclopedia.
Included index.
ISBN 1-85435-433-7
1. Habitat (Ecology) I. Marshall Cavendish Corporation.
II. Marshall Cavendish international wildlife encyclopedia.
OH541.W635 1992
591.5–dc20 91-35958
CIP

CONTENTS

Winter

Some birds actually flock to the deciduous woodlands from their breeding grounds in the north. The brambling, for example, invades woodland to feed, especially on beech nuts.

Many insects pass the winter as eggs, tiny larvae or pupae, well-hidden among the twigs and branches or buried under the leaf litter. There they are not only vulnerable to frost; they may also be threatened by animals that remain active during winter, such as shrews.

Hibernation

Many animals avoid the cold by sleeping through the winter. Mammals that hibernate enter a sluggish state, during which their bodily processes slow down to the minimum necessary for staying alive. For an animal such as the dormouse – a typical small mammal of the woodlands of western Europe – hibernation is a good solution to the rigors of winter, when food is less readily available than at other times of the year.

During the warm months, the dormouse feeds on nuts, fruits and berries, maybe storing a few in its nest for the winter. In the fall, it fattens itself up on acorns, chestnuts and hazelnuts, and when the temperature falls consistently below 60 F it prepares to hibernate.

The dormouse builds its nest on or under the ground, where it rolls up into a ball by tucking its head down onto its chest, curling its legs up to its chin, and clenching its feet. It then curls its tail up between its legs, over its face and over the top of its head. It also shuts its eyes and mouth tightly, and folds back its ears. The animal's muscles are clenched so tightly that it can be rolled along a flat surface, its whole body hard and cold to touch. By making itself into a tight ball, the dormouse minimizes the area exposed to the air, enabling it to keep its body temperature as high as possible.

The dormouse hibernates from September to April, although if the weather warms up enough, it will wake and go outside to feed for a while before returning to its nest. It is not only warmth that awakens the dormouse; if the temperature drops significantly below freezing, it is likely to die of cold. As a precaution, it wakes and becomes active until conditions improve, even though it expends a large amount of stored energy in doing so.

Although many birds fly south for the winter, surprisingly few bats migrate, even though they are the only mammals that can truly fly. In America, those bats that do migrate south for winter include the red, hoary and silver-haired bats. They travel down the Atlantic seaboard of the USA to spend the winter in the south-eastern states. Most bats, however, hibernate – often in caves where the temperature remains cool and constant throughout the cold months.

ABOVE Chaffinches are common woodland birds throughout much of Europe. They feed on seeds, such as those of cereals and weeds, but their stout, strong bills enable them to cope with larger seeds, such as beech nuts. Although chaffinches have distinct markings, they merge well with the background, whether they are feeding on the ground or among foliage. Their camouflage helps them to avoid predators such as sparrowhawks, which prey on small birds.

The long doze

Some animals avoid bad weather without going into a deep hibernation. The American black bear does not actually hibernate, but spends winter in a den where it goes into a long sleep. Its body temperature remains at a level that is close to normal, but its heartbeat slows down from 40 beats a minute to 10 or less.

During the long winter, the female black bear gives birth to one to three cubs, the number depending on the amount of food she was able to find in the fall. If the berry crop failed, she will not have put on enough weight to support both herself and suckling cubs through the cold months, and she will emerge from her winter den alone. If she has borne cubs, the family will emerge together in April or May, the cubs already about two months old. Having come through their first winter, the youngsters stay with their mother for between one and two years.

Spring flowers

With the arrival of spring, the deciduous woodland floor is covered with a carpet of flowers. By flowering early in the season, woodland plants

LEFT Flowers that grow on the woodland floor provide a splash of color in the shady green wood. Ferns, such as the "male fern" (top left), seen here with its orange spores, grow best in damp, shady conditions. Wild flowers of the European woodlands include: the white violet (top right); hepatica (center left); *Primula acaulis* (bottom left); and the sweet violet *Viola hirta odorata* (bottom right). The dog rose (center right) is a thorny shrub of the understorey with prickles that protect it from herbivores, although insect larvae feed on its foliage.

avoid the deep shade cast by the canopy as the year advances. Throughout the year, a variety of flowers can be found in bloom in the woods. Shade-tolerant plants bloom in the understorey, while others thrive at the woodland edge and in clearings.

In an oak and hornbeam wood, the amount of light reaching the ground in March has been measured at 52 per cent; this compares with 32 per cent in April, 6.4 per cent in May, and less than four per cent in June. The flowers, therefore, take advantage of the brief spring warmth and light to flourish before the canopy thickens and shades them out. As the plants spread their leaves, they soak up the sun's energy: building up reserves of nutrients that they store, often in underground bulbs, for the following year.

The wood at night

The woodland wildlife is almost as active on a summer's night as it is during the day. Although the loud chorus of birdsong is absent, the calls of a few night-time birds, such as owls or nightingales, echo through the trees. Much of the activity is concerned with feeding and hunting. Many caterpillars, including those of the purple hair-streak butterfly, feed at night, when insect-eating birds are roosting. However, they are still threatened by nocturnal animals such as mice and shrews. These small animals, in turn, are the targets of foxes and, above all, owls – hunters whose eyesight and hearing, powerful talons and virtually silent flight, make them deadly predators.

To escape daytime predators, moths take to the air during the hours of darkness. However, even at night they are vulnerable to predation from bats, which take the place of the swallows and swifts as the aerial insect-eaters of the night.

Woods and people

Humans have left their mark on most of the world's deciduous forests. Only in a few places, such as the Bialowieza forest of eastern Poland, does the woodland remain truly primeval – a genuine leftover from the ancient wildwood that

ABOVE The tawny owl is a woodland predator superbly adapted to nocturnal hunting. Its large, forward-facing eyes enable it to see well in the dark and judge distances accurately, and its hearing is extremely sharp. When hunting, the owl perches on a branch, watching and listening for prey. When it hears a rustle caused by a wood mouse or other small mammal, it glides down to seize the animal in its sharp talons, and carries it off in its short, curved bill. The owl's soft feathers enable it to fly silently, so that its victims do not hear it approaching, and so giving them no time to make their escape.

LEFT The hedgehog is one of the most well-defended animals in the woodland. Whenever it senses danger, or hears a sudden noise, it erects its stiff, sharp spines, which it possesses even as a juvenile. If confronted by a predator, it curls up into a ball to protect its underside, head and limbs. The hedgehog is mainly nocturnal, and has poor eyesight, although its senses of hearing and smell are good.

once streched across much of the European continent.

People made their first impact on woodlands when they began to clear it for agriculture. Finding the soils rich and productive, settlers cleared areas for planting crops. Tall, mature trees were cut down to provide logs for building boats and houses, falled wood and smaller branches were used for firewood, and grazing animals fed and browsed upon the woodland herbs and low branches. As settlements grew, so more woods were cleared away, including woods which harbored animals that were considered a threat to human life, such as wolves and bears.

Today, people are no longer so dependent upon traditional forest products, and traditional woodland management is dying out. In many parts of the world, this has enabled the woods to return to a wilder state.

The coppice

One of the most distinctive forms of woodland management is coppicing – a system that provides a regular supply of modest-sized timber for simple domestic items such as poles for fencing. In Europe, coppicing dates back to the mid-13th century. To produce a coppiced wood, the slender trunk of each understorey shrub is cut right back to a stump. The stump sprouts several straight trunks that can be harvested in due course. In order to ensure a harvest every year, different areas within a wood are cut annually. Within a single coppice wood, therefore, there will be open areas that have just been cut and denser areas that will be ready the next year. Among the coppice, some mature trees, known as standards, are allowed to grow to provide larger timber.

Coppicing provides a habitat that favors animals and plants typical of the understorey, rather than those that are associated with the tall trees of mature woodland. In open, recently coppiced areas, the light that floods the woodland floor encourages the growth of wild flowers and attracts butterflies, whereas the more dense coppice areas are ideal for small mammals and nesting birds.

Life on the edge

The clearance of great sweeps of deciduous forest which leaves behind small islands of woodland, has increased the amount of woodland edge. The results have benefited animals more typical of open ground and woodland fringes than of dense forest. In Europe, the kestrel and its American equivalent, the sparrowhawk (only distantly related to the European sparrowhawk), have adapted so well to the changes in their original deciduous woodland habitat that they now hunt along expressway verges and over urban wasteland.

Woods in towns

Many modern towns and cities are built on land where forests once stood. Although the ancient wildwood has long gone, trees remain an important part of city life. They are planted along avenues, and provide welcome shade in city squares and gardens. Where land is abandoned, it is eventually colonized by weeds, grasses, shrubs and trees, so that woodland may reappear in areas where it once grew naturally.

Urban woodland contains trees that can survive in the disturbed, often dry, and sometimes polluted, soils of towns. In London, the most common species of tree include sycamore, ash and birch. Although the wildlife that inhabits the urban woods is limited to the bolder woodland animals, the habitat still provides homes for a wide range of species.

Feast for warblers

The large numbers of invertebrates that live on sycamores during the spring and summer are an ideal food source for nesting warblers. Small rodents are abundant under the trees, and where the canopy is open, these creatures attract the attention of hovering kestrels. The fox has also become an extremely successful urban animal in large towns and cities.

ABOVE The nightingale is a species of thrush that lives in deep cover within shrubby vegetation. It is a drab bird, and its lack of distinct markings make it difficult to see among the leaf litter and shady thickets. Although the male uses his reddish tail in his courtship display, it is his song that acts as the main form of communication in courtship and in defining his territory. Nightingales feed on the ground, where they hunt for invertebrates, from large earthworms to tiny insects and spiders. In the late summer and fall, they eat berries too. Their strong legs are an adaptation to ground level feeding, and their large eyes enable them to search efficiently in the dim light of the woods.

ABOVE The chiffchaff is a typical Old World warbler, having a finely pointed bill that is ideal for capturing insects. Although a small bird, weighing less than 0.5 oz, the chiffchaff is a skilled and strong flier. It catches insects on the wing and migrates great distances between its breeding and wintering grounds. Its strong feet are well-suited for perching, climbing on thin twigs and moving through rough undergrowth.

European woodlands

The deciduous woodlands of Europe all contain the same species of trees, although their combination varies from place to place. Beech and oak are the dominant species, while hornbeam, chestnut, ash, sycamore, hazel, field maple and lime also occur.

European woodland is a rich habitat for birds. Some species, such as warblers, blackbirds, woodpeckers and tits, feed and breed in the woods. Others, such as wood pigeons, grey herons and rooks, breed or roost in the woods, but feed elsewhere. Woodlands are alive with birdsong from the beginning of spring, when the warblers return from their wintering grounds in the south. The numbers of returning migrants are depleted by the stresses of their long journeys. On reaching their winter homes, they have to adapt rapidly to their new habitat, and some inevitably fall prey to predators.

Unfortunately, those birds that survive the winter face great danger when traveling over southern Europe, since human hunters in this region shoot any creature out of the sky in the name of sport. Perhaps in the days of falconry and crossbows, the killing might have helped to eliminate the weakest birds, leaving the strongest stock to breed. But in the age of highly accurate guns, it simply results in the wholesale slaughter of birds every spring and fall.

Dull-colored but tuneful

Warblers are dull in color, which helps to camouflage them from predators in their densely vegetated habitat. But because the different species are so similar in color, they cannot rely on visual recognition in courtship and mating. Instead, the warblers rely on their calls to attract mates, and many have strong and distinctive songs.

Two common European warblers, the willow warbler and the chiffchaff, are almost identical in appearance, both being a dull brown. However, their songs are different, ensuring that females rarely mistake a male of the wrong species for a potential mate. The willow warbler has a song consisting of a series of single notes, falling down a scale. The chiffchaff's song resembles the tonal change of its name, being two similar notes – chiff-chaff – that are frequently repeated. Although these two species of warblers have similar coloration, they have quite different mating behavior.

Feathered harvesters

Warblers are insect hunters. They have strong, gripping feet that enable them to move deftly through the twigs and branches, and fine, pointed bills well-suited for plucking their victims off the leaves and bark. Along with flycatchers, tits and finches, warblers are important harvesters of the invertebrates of the woodland.

Flycatchers have wide beaks that enable them to catch flying insects, and sensory bristles around their beaks that assist them in detecting prey at close range. They are not

BENEFITS FOR BIRDS AND PLANTS

Some animal and plant species in the deciduous woodlands have evolved symbiotic relationships – relationships that benefit them both. The mistle thrush, for example, greatly benefits the mistletoe – one of the plants on which it feeds and from which it takes its name. Mistletoe is a parasitic plant that grows on the branches of apple trees, among others. The mistle thrush feeds on the berries of the mistletoe, but is unable to digest the hard little seeds, which pass through the bird and are expelled in its droppings on the branch. Alternatively, the seeds stick to the bird's bill and are later wiped off on a branch. In this way, the mistletoe's seeds are transported to new sites.

Transporting acorns

The jay helps the seed dispersal of the oak in a similar way. As fall turns into winter, the jay gathers acorns. It hides them away in the wood, under leaves or moss or in holes that it digs with its beak. The jay has an excellent memory for its hiding places, returning to them to recover the food throughout the winter, even locating them under snow. It also digs up acorns that have germinated in spring, having identified the young seedlings. Only a handful of acorns have to be forgotten by the jay, or seedlings ignored, for a new generation of oaks to germinate.

RIGHT The jay, a colorful member of the crow family, is especially active during the summer months when it buries acorns and other seeds in caches to supplement its winter diet. Jays are often responsible for burying acorns far from their parent tree; if the birds do not find them again, the acorns may well germinate and grow into infant oaks.

particularly adept at maneuvering on foot among the twigs and branches, but they are fast and agile when giving chase to an insect. They sit on a prominent perch and dart out to capture insects on the wing. Spotted flycatchers can catch about three insects a minute in suitably warm, still conditions during summer; they remove the sting from bees and wasps before eating them.

The chaffinch is mainly vegetarian. It feeds on the seeds of a variety of plants through much of the year, and supplements its diet with insects during the summer. These are a useful food source when the seed supply dwindles, and contain a high protein level that is ideal for rearing the chaffinch's young.

Different niches

The ecosystem in woodlands can support a high number of different

RIGHT Oak trees are important for a great variety of woodland wildlife – invertebrates feed on the leaves and wood, birds nest among the branches and in holes in the trunk, while various birds and mammals feed on the acorns.
FAR RIGHT The 450 or so species of oaks in the world's temperate woodlands can be recognized by the shapes of their leaves and acorns. The leaves may differ within the same species according to habitat – the leaves of an oak that grows in the sun are more deeply lobed or smaller than those that grow in the shade – an adaptation that reduces the amount of leaf surface exposed to overheating (both variants are shown here). Among the European oaks are: the Turkey oak (1, 1a, 1b); the durmast oak (3, 3a, 3b); and the Hungarian oak (4, 4a, 4b). North American oaks include: the scarlet oak (2, 2a, 2b); the water oak (5, 5a, 5b); and the canvon live oak (6, 6a).

1
1a
1b
2
2a
2b
3
3a
3b
4
4a
4b
5
5a
5b
6
6a

LEFT The great spotted woodpecker is specially adapted to climbing trees. Sharp, curved claws enable it to grip the vertical surface; two of the toes point forwards and two back, so that they can spread out to grip onto any trunk or branch, regardless of its contours. In addition, stiff tail feathers support the bird while it pecks at the wood.
FAR RIGHT Great tits are common woodland birds in Europe. Their young sit motionless in the nest when disturbed, relying on their moss-green backs and the broken color on their heads to provide camouflage. Only the female great tit incubates the eggs, and each brood may contain up to 16 nestlings.

bird species provided each has its own particular niche in diet and behavior. Although the numerous bird species that inhabit the deciduous woods may share the same general habitat, they differ in certain characteristics, such as feeding methods, the exact type of food, and the location of feeding sites. Such distinctions between the species ensure that there is enough food to go round.

The males of the various warbler species sing to communicate with the opposite sex and to define their territory. Each male has a well-defined area, which it defends against rivals. The territory contains shelter, the nest and the source of enough food for itself, its mate and offspring.

Chiffchaffs and blackcaps

The blackcap is a warbler that inhabits the European deciduous woodlands, often occurring in the same sites as the chiffchaff. The two birds have slightly different feeding habits, which reduces competition between them, even though they both feed on the small invertebrates that live among twigs and leaves. Chiffchaffs either dart out from perches to catch flying insects, or hover to pick them off leaves. The blackcap rarely flies after its prey, preferring to take larger insects that have settled on twigs or leaves.

Tits are among the most familiar of woodland birds in Europe. Several similar species occupy slightly different niches within the same habitat. The blue tit is light and agile, and feeds on insects on thin, swaying branches. The great tit is more bulky and has a larger bill. It finds its food on the thicker, lower branches of shrubs.

In a hard winter, when snow is on the ground, the blue tit picks beech nuts off the tree by climbing along the thin twigs. The great tit, which is more dependent on nuts lying on the ground, is unable to reach those nuts still on the tree, and may starve.

Woodpeckers are closely related to the warblers, and also feed in different ways. The lesser spotted woodpecker hunts high in the canopy; the greater spotted woodpecker feeds among the main branches and on the trunk; and the middle spotted woodpecker finds food on dead trees. In spite of having precise niches, the birds still encounter conflict. The great tit and blue tit, for instance, compete aggressively for nesting sites, with the great tit generally proving the victor. Both great and blue tits attack and rob food from members of their own species in winter, and both attack smaller tits that share their habitat.

Invisible mole hills

The European mole finds the fine soils of woodlands ideal for digging its tunnels. There is usually little evidence of a mole's presence in the woodland, since mole hills tend to be well-concealed by scrub and undergrowth. But they are usually present in large numbers, feeding on the abundant supply of earthworms.

In the winter months, when the earthworms bury themselves deep down in the soil, and the ground is possibly frozen, the mole may face a serious food shortage. To prevent this, it lays up stocks of food in the fall. It may collect hundreds of earthworms, immobilizing each one with a bite behind the head, and leaving it in the tunnel to be eaten later. If a mole leaves an earthworm alone for long enough, the worm's ability to regenerate itself may enable it to recover and creep away.

The wood mouse lives among the plants and bushes above the mole's tunnel. It is a common woodland rodent that is active at night. It feeds mainly on seeds and nuts, but it also eats invertebrates. The wood mouse is an important source of food for

ABOVE The pygmy shrew has a long, mobile snout and sensitive whiskers for locating the insects on which it feeds. As with all shrews, the pygmy shrew needs to eat frequently to replace body energy lost through heat radiation. As a result, it moves restlessly day and night through low vegetation in search of edible invertebrates, including snails and worms.

two of the most effective predators of the wood; the tawny owl and the weasel.

During the day, the tawny owl hides among the branches, close up to the tree trunk. Its dull coloration camouflages it well against the bark, but if small birds discover it during the day, they will heckle and mob it, sometimes even forcing it to find another perch. At night, the owl is in its element, a silent hunter that swoops down to kill its prey of small animals, and occasionally birds.

The goshawk is a large bird of prey that hunts in the daylight. Its main target is the sparrowhawk, to which it is related, but it also captures pheasants, one of the most common game birds. As a result, the goshawk has been persecuted in those parts of Europe where the pheasant is reared for shooting. A native of Asia that was introduced to Europe centuries ago, the pheasant is shot in large numbers during the winter season.

In many places, pheasants are reared in pens as young birds before being released into the wild. Gamekeepers encourage them to stay in the wood by putting down grain. They also keep down the numbers of predator that prey on pheasants or other game birds. The gamekeepers kill foxes, birds of prey, weasels, polecats and magpies because of their reputations for taking either eggs, young or adult game birds. In some places, goshawks have learned to hunt in the rearing woods, where the pheasants are easy prey. In Finland, more than 5,000 goshawks are killed each year by game hunters protecting their pheasants. Despite this persecution, the goshawks have managed to survive in fair numbers.

Non-stop shrew

Common shrews usually inhabit the thick, scrubby areas of woodland, and they also occur in hedgerows. They have a rapid population turnover. Breeding in the spring and summer, they produce several broods, each consisting of five or more young. Despite this increase in numbers over the warmer months, shrew populations remain more or less constant from year to year – their numbers being controlled by competition for food and by the weather.

A shrew lives for about 18 months. During the fall when invertebrate populations, depleted by predation throughout the year, get smaller and smaller, the youngsters and their parents compete for diminishing food supplies. The older shrews, unable to compete successfully with their offspring, die, enabling the new generation to thrive.

Shrews are active animals that need to eat about three quarters of their own body weight in food a day. Consequently, they spend most of the time hunting, and only rest for short periods. They are efficient hunters that dig into the woodland soil to find food as diverse as bulky pupae and tiny worms.

In winter, many young shrews die of the cold. Only the strongest survive until the spring, with its plentiful supply of food. They then feed and breed – until another generation of young replaces them in the autumn. Although their flesh is distasteful to most animals, shrews

are hunted by tawny owls. When the shrews die in the autumn, their bodies usually remain lying on the ground, ignored even by species that normally feed on carrion.

The fire salamander is less active than the shrew. It lives in damp places, usually close to water, where it feeds on invertebrates, hunting them at night as it slowly patrols a small area around its hideaway. The black and bright-yellow body coloring of the fire salamander warns predators of the foul secretions that its skin produces. The secretions irritate the eyes and mouth of any animal that attempts to eat it.

The hedgehog, a relative of the shrew, has sharp spines on its body that serve as an effective defense against predators. Unlike the shrew, it hibernates through the winter, huddled under the cover of the thick leaf litter.

ABOVE The lime hawk-moth belongs to a large family that ranges across Europe to Japan. It is especially common in towns and cities where lime is planted as a street tree. The lime hawk-moth's larva feeds on the leaves of a range of deciduous trees, but it prefers lime. Like most hawk-moth larvae, it has a stiff spine at the rear end of its body that may help to deter predators. Hawk-moths have stout bodies and long, narrow wings that make them among the strongest fliers of all the butterflies and moths.

Bats Versus Moths

To protect themselves from predators, many woodland insects have evolved defenses such as natural camouflage and repellent secretions. Most moth species, however, can offer little resistance when attacked by one of the most ferocious of the nocturnal insecteaters – the bat.

In the open air, a moth has nowhere to hide from a hunting bat. The bat locates its prey by emitting ultrasonic pulses that bounce back from objects in the area. As the bat flies, it picks up ultrasonic echoes and uses them to work out a mental picture of its surroundings. Its advanced used of sound allows it to navigate, locate insects in the dark and catch them in mid-flight. Its agility usually guarantees a catch.

The horseshoe bat is one of the most agile bat species: in less than a second, it can analyze the exact pattern of a cave roof in complete darkness, choose a tiny nick or lump on the roof as a perch, turn a somersault in the air and hang upside-down.

Early warning systems

Many moth species are defenseless in the face of the speed and efficiency of predatory bats. Some families, however, have simple ears that pick up the ultrasonic pulses emitted by bats. As soon as such a moth detects an approaching bat, it performs speedy evasive maneuvers or dives for cover.

Some moth species have the ability to jam bat signals. The American banded tussock moth emits a sequence of ultrasonic clicks when it hears a bat approaching. When confronted with this interference, the bat becomes disoriented and swerves away. A few bat species foil the signal-jamming moths by emitting sound pulses that fall outside the frequency range of the moth's hearing.

BELOW The ultrasonic pulses that the brown long-eared bat emits to locate its insect prey are very quiet, suggesting that the bat can capture moths using its sense of hearing. The bat's large ears enable it to pick up the faint returning echoes of its own pulses. As the pulses bounce off solid objects, the bat can analyze the layout of its environment in great detail; it can even detect and catch small insects that have settled on leaves.

Understorey animals

Dormice are typical mammals of the European woodlands understorey. They have strong, slender, flexible feet that enable them to grip thin branches as they clamber around in search of nuts and fruit. Their nests are well-hidden among the branches and are built of shredded honeysuckle bark.

The robin, one of the best-known woodland birds of western Europe, feeds on insects that it finds among the leaves or on the ground. In a mature deciduous wood, where plenty of food is available, more than 200 pairs of robins may inhabit each square mile. Human management of the woodlands has resulted in changes that have greatly benefited the robin. Being an understorey bird, and an adaptable feeder, it has successfully moved from its original woodland haunts to take advantage of hedgerows, gardens, the edges of woods, and new habitats created by coppicing.

The nightingale and the woodcock are among the few nocturnal bird species to live in the European deciduous woodlands. The woodcock is the only woodland wader; it probes for earthworms in the damp soil with its long, slender bill.

A coppiced home

The traditional cycle of coppicing in woodlands has created a habitat that suits the nightingale. Since the bird likes a habitat in which many thickets grow, it prefers coppices that are about five years old (in which the trees are roughly twelve feet high). In sites where coppicing continues, the nightingale thrives and its night-time song is a distinctive feature. Unlike the robin, the nightingale has not colonized similar shrubby habitats outside the woods.

The holly and the ivy

Ivy is one of several species of climbing plants that thread their way up from the ground towards the canopy, taking advantage of the trees to reach the light. Both ivy and holly are the food plants of the holly blue butterfly. In spring, the female lays her eggs on the male and female holly flower buds. The eggs laid on male buds die, because the newly hatched caterpillars depend upon developing berries, which are only produced by the female buds. Why the butterflies waste so many eggs remains unknown. The adults that develop from the spring eggs produce a second brood that feed on ivy.

ABOVE The larvae of the oak processionary moth spend the day in a communal tent of silk that they make in the branches of an oak tree. At night, the larvae emerge to feed on oak leaves. They form a line as they travel through the branches or along the woodland floor, each larva following the one in front. After a night's feeding, the group returns to its nest along the same trail.

To bee or not to bee

The caterpillars of the broad-bordered bee hawk-moth feed on another climbing plant, the honeysuckle. The adult moth, a large day-flying creature, resembles a bumble bee (especially in flight) – a form of mimicry that discourages potential predators from eating it. Many woodland insects use mimicry to warn would-be predators that they are poisonous, distasteful or dangerous.

Graceful deer

Red, roe and fallow deer graze in the European woodlands in great numbers. They were once hunted by woodland carnivores, such as the wolf, but now that the European forests are free of large predators, the deer feed in peace. But their numbers have to be kept down to prevent them from damaging forest trees by browsing or stripping bark from young trees in the rutting season.

The rut, which occurs in late summer or early autumn, is a noisy affair in which the males dispute aggressively over females, and fight each other for the right to mate with them. The males use their antlers in these disputes to thrash vegetation or to clash against those of their rivals in duels. Once the rut is over, the male deer shed their old antlers and grow a new set for the next year's rut. Antlers are often damaged in fights, and because their size affects the male's dominance over rivals, it is vital that he grows a new set if he is to be successful for more than one year.

The young deer have white-spotted, rust-red coats that provide them with perfect camouflage as they lie hidden among the bracken and other vegetation on the sun-dappled woodland floor.

The bison was once the greatest herbivore of the European woods, but was hunted to extinction in the wild in the early years of this century. Fortunately, a number of individuals remained in captivity, and the bison was re-introduced to the forests of Bialowieza in Poland, the greatest remaining stretch of virgin woodland in Europe. Bialowieza has remained unchanged since the days of the great wildwood, when deciduous forests covered most of Europe before the last Ice Age.

ABOVE When male red deer fight over females, they lock their antlers together and twist and turn in an attempt to push each other aside. The deer are easily hurt; on one Scottish island, nearly a quarter of the stags were injured during the rut each year, some of them suffering permanent damage.

RIGHT The fallow deer is well adapted to avoiding predators. Its speckled coat provides superb camouflage in the dappled woodland light, and its large eyes – placed on the sides of the head – give the deer good all-round vision. While grazing, the deer will suddenly lift its head and look around for any threat, at the same time turning its large, sensitive ears in the direction of suspicious sounds.

North American woodlands

The deciduous woodlands of North America contain a far richer variety of plant species than their counterparts in Europe. One theory suggests that is a result of the differing alignment of mountain ranges in America and Europe. The mountains in the New World lie on a north-south axis, whereas those of Europe run from east to west. In the last Ice Age, about 10,000 years ago, the flora of North America stayed ahead of the advancing ice by moving southwards. In this way, it retained its richness of variety. When the climate warmed and the ice retreated, plants were once more able to grow in areas further north. In Europe, however, the

ABOVE Like all the members of the weasel family, which also includes otters and badgers, the striped skunk has a pair of anal glands that produces a foul-smelling substance known as musk. The skunk uses its musk to mark out its home range, and to defend itself. If the skunk's conspicuous black-and-white coat fails to deter other animals from molesting it, it first raises its tail over its head and stamps its feet; it then discharges the contents of its anal gland, directing the yellow spray with great accuracy at the aggressor. The musk is not lethal, but it is pungent enough to drive any animal away. At night, striped skunks are at their most active, foraging about in their territory for food.

LEFT Swamp cypresses are conifers that grow up to 150 ft high and are typical of the damp woodland habitats favored by striped skunks.

species were forced back against the mountains. They had nowhere further to retreat, and those that could not adapt to the conditions died out.

When European settlers arrived in America, they found many plant species similar to those that grew in Europe. The American woodlands feature combinations of oak and hickory, oak and chestnut, beech and maple, and many other mixtures. Other species include basswood and tulip trees.

The animal species that occur in the North American deciduous woodlands are rarely related to those that inhabit the European woodlands, yet they have many similarities and occupy similar habitats. For example, the New World wood warblers of North America have filled the same niche as the European warblers; and the deer mouse of America fills the same niche as the wood mouse of Europe, although the two species are not related. Animals of this kind are known as "equivalents."

A convergent evolution

When habitats, with similar niches, are separated by large distances or insurmountable barriers, unrelated plant and animal species often evolve similar adaptations – a process known as convergent evolution. Some of the equivalents however, are related: the North American equivalent of Europe's red deer is the wapiti (also known as the American elk), and the two are sometimes considered to belong to the same species. Both the male wapiti and the red deer stag are extremely noisy and aggressive during the rutting season, when they

ABOVE **The gray fox is a retiring, largely nocturnal animal that occurs in forests and brushland from southern Canada through the USA and Mexico to northern South America. Unlike many other foxes, it is an agile tree-climber, using its strong, short legs to climb into the branches either to escape enemies or to search for fruit in season. Other food in its varied diet includes birds, mammals (especially rodents) and insects.**

fight with other males to secure females for mating.

Continued conflict

The deep roars of the wapiti stags do not represent the only conflict in the North American woodlands. The herbivorous and carnivorous insects of the canopy live in a constant state of conflict with each other. If conditions are particularly good for a herbivorous insect species one year, it thrives and multiplies. In turn, the population of its carnivorous enemies increases dramatically. As they feed, they reduce the numbers of their herbivorous prey. When their food is scarce, the population of the predator species decreases, leaving the herbivorous prey to multiply again. The constant rise and fall of these populations is a vital part of the balance of woodland wildlife.

The relationship between predator and prey is of great economic importance in temperate woodlands. In Canada, in 1977, caterpillars of the forest tent moth completely defoliated an area as large as Texas. By stripping the leaves of broad-leaved trees, including oak, sugar maple and trembling aspen, the caterpillars reduced the growth of the trees, making them weak and prone to disease. Over 2.7 billion cubic feet of wood was lost, as well as vast amounts of sugar that would have been harvested from the sugar maples if they had been in leaf and producing sap at a healthy rate.

Pests often stop short of killing host trees to ensure that they do not leave themselves without food for the next season. The forest tent moth was brought under control by the larvae of the flesh fly. The female flesh fly lays her eggs inside the cocoon of the forest tent fly's pupa, and the growing flesh fly larva eats the developing pupa. The flesh fly acts as an important biological control on the forest tent caterpillar, and has considerable economic benefit to the forester.

North American marsupial

The Virginia opossum, a widespread mammal the size of a small cat, is North America's only marsupial. It occurs in woodlands and

ABOVE The wood duck of North America lives in wet woodlands, where it nests in hollows in the trees and perches on the branches with the aid of its strong, sharp claws – an ability possessed by few waterfowl. The duck's hind toes are well developed and provide support when the bird is climbing, while the short broad wings enable it to maneuver with agility as it threads its way through the woodland trees.

other habitats, where it forages on the ground and in the trees for animal and vegetable food. The young are born early in the year, and develop in their mother's pouch. They are so small when born that a litter of up to 14 babies can fit into a teaspoon.

Winter in America

Animals of North America have adopted various techniques to ensure that they survive the bitter winter months. The striped skunk, for example, hibernates in a den. The red-headed woodpecker migrates south until the following spring, while the hairy woodpecker changes its diet. During the breeding season, it feeds on invertebrates, but in the winter months when they are scarce, it eats seeds.

Bobcat territory

The most widespread cat of the temperate parts of North America is the bobcat, which inhabits areas of deciduous forest. It feeds on small mammals and birds and stands at the top of the forest food chain. The bobcat is a territorial animal that marks out its home range with scent and marks scratched on tree trunks. It may extend its territory, neighbors permitting, during the course of the year if food supplies run low. The home ranges of male bobcats do not overlap with each other, but a male's range may often contain the ranges of several females. The males and females tolerate the competition for food that results, in exchange for finding a mate.

Woods of the Far East

Eastern Asia contains both deciduous and evergreen temperate forests. Unfortunately, large parts of these woods have been felled, and in many places, the forests only remain high in the mountains. A rich variety of tree species grow in these woodland areas, and many of them are related to those in European and American woodlands. They include oaks, birches and maples, as well as various species of pine.

As with other deciduous woodlands, the woods of eastern Asia support large populations of small mammals and birds. These provide food for a range of predators, such as the racoon dog. Unrelated to the racoon, but similar in appearance, the racoon dog eats hedgehogs, shrews, grass snakes, insects and ground-nesting birds such as larks and pheasants. It is particularly good at catching fish and amphibians in streams. The racoon dog sleeps heavily during the cold weather, but it emerges from its den on warmer days.

The Himalayan black bear is active during winter, only retiring to its den in particularly cold conditions. Widespread throughout the forests of eastern Asia, it climbs well, easily scrambling up into trees where it makes a nest of sticks in which to rest during the hot weather. It is an omnivore, raiding bees' nests for grubs and honey, as well as feeding on nuts and fruit.

Feline hunters

The temperate woodlands of eastern Asia support leopards and tigers, both of which feed exclusively on meat. The powerful tiger hunts and kills large prey such as deer, while the leopard usually feeds on smaller animals, often dragging its victims into trees to keep them from scavengers.

Introduced Species

The similar ecology of temperate woodland ecosystems throughout the world has made it relatively easy for animals from one part of the world to colonize another. When animals are introduced to new habitats by humans, they may have a disruptive effect on a previously stable ecosystem: animals that enter new locations, accidentally or otherwise, do not always mix well with the native inhabitants.

Gray squirrels were introduced from North America to Europe, where they adapted successfully to their new conditions. Unfortunately, they were so successful that they displaced much of the red squirrel population in native deciduous woods. Gray squirrels also inhabit towns, where they are far from shy and attract considerable attention.

Sika deer were successfully introduced from Japan into European woodlands without causing serious damage to the environment. However, like all deer in Europe, their population may become too large for the habitat, since there are few large predators to control their numbers.

Causing havoc in New Zealand

New Zealand's only native mammals are two species of bat – in the absence of predators, both species have become flightless. Birds evolved to fill the niches that are occupied by mammals in other woodlands.

Humans have created havoc by introducing 20 different mammal species from around the world. Domestic cats have destroyed many flightless birds and heavy grazing by introduced herbivores, such as red deer, prevents forest regeneration.

One positive outcome of the New Zealand introductions is the preservation of the parma wallaby – a species that was introduced from Australia, but which later became extinct in its native land. It has since been reintroduced from New Zealand stock.

BELOW The gray squirrel is a skilled climber, having sharp claws for gripping, and long toes for holding onto the uneven surfaces of tree trunks. It is also built for leaping: its bones are light, and when the animal jumps from trunk to trunk, the loose skin on its flanks spreads out like a parachute, preventing it from falling too quickly. As it leaps, the squirrel fluffs out its tail, gaining extra lift.

Hot spring dippers

One of the most remarkable occupants of the northern Japanese forests is the Japanese macaque. The largest of the macaques, it has adapted to survive the freezing cold winters. It has a thick winter coat, and is renowned for its habit of bathing in the hot springs to keep warm. Despite being a forest animal, the Japanese macaque spends most of its time on the ground where it feeds on leaves, crops, fruit, insects and small mammals. In winter, it finds enough food under the snow to prevent it from starving. It also forages along the seashore for shellfish.

Moist evergreen woods

The temperate evergreen forests of the west coast of America are different from any others. The forests are dominated by huge redwood trees – the largest trees in the world, often measuring over 300 ft high. Other conifers, including Douglas fir, Sitka spruce and hemlock replace the redwoods towards the north. More than 15 in of rain falls each year in these west coast forests, and as a result of this high level of moisture, moss grows in abundance and water drips continually from the foliage.

The wapiti is a common inhabitant of the west coast forests, which in the fall are filled with the bellowing of the wapiti stags in rut. Other animals that inhabit the western forests include garter snakes and black bears.

The puma stands at the top of the food chain in the evergreen forest. Like all wild cats, it is a shy and secretive hunter. Growing up to six feet long and weighing up to 170 lbs, it preys mainly on the wapiti, often catching it by pouncing from a branch. The wapiti is a fast runner that can evade a chasing puma, and so ambush techniques are useful to the cat. Although the puma is well-suited to the forest, it also occurs in more open, arid habitats.

The wet forests of Florida

The evergreen woodlands of the south-eastern United States contain a variety of tree species, including evergreen oaks and several types of pine. Some areas of the woodlands are wet, and the conditions are almost subtropical.

A variety of animals live among the trees, including the Virginia opossum, the white-tailed deer, the black beer and the gray fox – animals that also inhabit the deciduous woods to the north-east. A number of species live only in the warm, wet, subtropical habitats. These include a variety of turtles living in the wooded swamps, tree-frogs that eat invertebrates in the trees, and zebra butterflies that flutter through the branches.

Land of flightless birds

The forests of New Zealand's North Island are subtropical, whereas those of South Island are temperate. The most dominant trees are the four species of evergreen southern beeches that thrive in the wet conditions and have moss hanging from their branches.

A range of flightless birds inhabit both islands. Those that occur in the South Island beech forests include the takahe and the kakapo. The takahe nests high in the mountains, moving to the beech forests for the winter. The kea is the only parrot species that eats meat.

ABOVE Although the Siberian blue robin, a member of the thrush family, often perches in trees, it forages mainly on the ground. Like many ground-feeding birds, it has long, strong legs. It occurs in both coniferous and mixed woodlands, and haunts areas of dense undergrowth.

Habitat under threat

Temperate woodlands often occur in areas with high human populations. The demand for land for housing and cultivation has led to a severe reduction in the habitat for the wildlife, while hunting has threatened many species with extinction. The European bison was saved from extinction at the eleventh hour; the wild boar, hunted almost to extinction in Hungary earlier this century, is now confined to nature reserves. Several races of sika deer in Asia are now facing extinction through loss of habitat. Only by retaining large areas of woodland – keeping some of it wild, while some of it is managed in a traditional way – will the future of woodland fauna and flora be assured.

HEATHS AND MOORS

The purple-flowered heathlands, and the lonely, upland moors, are fragile and fragmented habitats where nightjars churr hauntingly at dusk, dragonflies dart after prey, and wall lizards soak up the sun

PAGES 228–229 Heaths are shrubby habitats with few or no trees. Their soils are poor, acid and low in the nitrogen that plants need to make proteins. However, several heathland plants have adapted to these difficult conditions and obtain their nitrogen from alternative sources. For example, gorse and broom (seen here) have bacteria in their roots that fix atmospheric nitrogen and build it up into proteins.
ABOVE The Dartford warbler is present throughout the year in parts of continental Europe and southern Britain. It is the only British bird that is entirely restricted to heathland, where it lives among the dense gorse and long heather.

Lowland heaths and their upland equivalents, the moors, do not contain the diversity of wildlife that is typical of other temperate habitats, but they do contain a distinctive array of animals and plants. Of the low, shrubby vegetation that covers heaths and moors, the most familiar is heather, which brings to these regions the characteristically soft, purple colors of summer. There are other low shrubs, many of which are related to heather. In places, the heather-dominated heath grades into grassy heath or wet, spongy ground known as bog.

Heathland animals include colorful butterflies and moths, numerous spiders and several species of birds. In summer, snakes, such as vipers, are occasionally seen basking on a sunny bank or sliding into the thick heather if they have been disturbed.

Origin of heaths

Most heathland owes its existence to human activity. It develops when woods are chopped down for agriculture, and the land is maintained as low scrub for grazing animals or game birds. In some coastal and hilly sites, heath does occur naturally, probably because the strong and persistent winds make it impossible for trees to survive.

"True" heathland occurs in north-western Europe, especially in the British Isles. Scrubby habitats elsewhere in the world are sometimes defined as heath, especially those of South Africa (where it is called *fynbos*, pronounced "fain-boss"), and parts of Australia. But they are different in many ways to the heaths of north-western Europe, which are sometimes called Atlantic heaths. Small areas of heathland also occur in North America. Some heaths are found above the treeline on mountains in certain parts of the world, but these are best regarded as mountain habitats.

Heathland conditions

In north-western Europe, especially near the Atlantic coast, heathland soils are acidic and poor, and the climate rainy and mild. Along the coasts, rain falls throughout the year, and even in the driest months, the ground receives at least 2 inches of rain. The average temperature for the coldest month is above freezing, but remains below 45F. The average temperature of the warmest month is above 68F but below 72F.

Heather is well-adapted to this type of climate. Being an evergreen, it can photosynthesize (use the sun's energy to produce chemical components) throughout the year, except when temperatures are very low. The plant's leaves are tiny and have a thick, waxy covering to reduce loss of moisture. Although this precaution may seem surprising in a wet climate, it is necessary in cold winters. When the ground freezes, the plant cannot draw water up through its roots and cannot, therefore, afford to lose moisture through its leaves. The heather's watertight leaves also guard the plant against drought, since in summer, the sandy soils of many heaths drain quickly, becoming very dry.

For heather to be the dominant heathland vegetation, trees must not be allowed to grow. In the past, humans burnt the heaths and grazed their animals on the land to ensure that trees did not grow. Heather is not affected by fire, regenerating quickly either from root-stocks left untouched in the ground after the burning, or from seeds. It produces large numbers of seeds in the fall and those that do not germinate immediately remain in the soil

RIGHT The purple flowers and dark green foliage of heather often dominate large expanses of heathland. Heather supports a host of animals and about 40 species of insect depend upon it for food. Spiders find the heather's dense branches ideal places for building their webs, while the hot, dry, open spaces amid the heather attract other invertebrates, including ants, wasps and butterflies.

until favorable conditions occur. So prolific is heather, that in one count more than 8,000 buried seeds were found in one square metre of land.

Heathland plants

British heaths are typical of the European heathlands where heather is the most typical plant. However, many other species also occur, including relatives of heather, such as bell-heather, cross-leaved heath and berry-bearing shrubs (bilberry and cranberry). Altogether, about 12 members of the heather family occur on British heaths. Two other familiar heathland plants are bracken and gorse.

Lichens are a characteristic feature of heathland. They grow on the ground below the heather, and on the stems of the heather itself. The lichens form a colorful, miniature forest below the heather canopy.

Defense against grazing

Lowland heath has a warmer and often sunnier climate than upland heath, and tends to be richer in invertebrates, although some species occur in both habitats. Many heathland invertebrates feed on the leaves of heather. They, in turn, are food for a variety of other invertebrates and larger animals. Compared with other habitats, heaths do not

ABOVE The flowers of heather attract bees, which take pollen and nectar and pollinate the flowers. The honey that honey bees make from the heather flowers is heavy, golden and fragrant. The different bee species have different methods of pollinating flowers, depending on the length of their tongues. The tongues of some species of bumble bee are too short to reach right down inside the flower; instead, they bite holes in the base of the flower to reach the nectar.

have a great range of leaf-eating invertebrates; the poor soils and the weather conditions mean that the plants are slow-growing and poor in nutrients. As a defense mechanism, the plants build up chemicals, such

as resins, in their leaves, that make them distasteful to many grazing animals – and inflammable in dry weather.

A touching relationship

Heather and the heather thrip have a relationship that benefits both the plant and the insect. The grubs of the heather thrip live inside the plant's flowers, feeding on their pollen. Although the flower loses vital pollen grains that are necessary for fertilizing other heather flowers, the plant gains in the long run. When the female grub reaches maturity, she forces her way out of the flower past the stamens, which brush pollen onto her body. After mating, the thrip visits other flowers to lay her eggs, where the pollen rubs off onto the flowers, pollinating them.

Heathland bugs include both herbivores and carnivores. Some of the former feed only on heather; others have their own specific plants. The gorse shieldbug, for example, feeds only on gorse. Predatory bugs include the heath assassin bug, which eats a variety of insects and their grubs. Mites also feed on gorse, and during summer, one species covers the bushes with its webs to hide the tiny animals while they feed. Some heathland herbivores, such as the heath grasshopper, are grass-feeders.

ABOVE Thyme is a wild flower of warm, dry places. Although it is not a typical heathland plant – it is more likely to be found in chalky habitats – it sometimes grows on heaths where chalk soils mingle with the heathland sands. Its tiny leaves are an adaptation to dry conditions; their small surface area keeps water loss to a minimum.

Although heath does not consist of complex layers, its surface cover varies in appearance. In areas that have recently been burned, bright green shoots emerge from the charred earth; in other areas, the vegetation may be long-established, tall and thick. The land can have dry, sandy patches or wetter, boggy spots. Each of these habitat varia-

LEFT The colorful wings of the male silver-studded blue butterfly have probably evolved as a means of attracting a mate. However, the butterfly runs the risk of also attracting predators such as insect-eating birds. To avoid them, the butterfly often rests with its wings closed, exposing the speckled and less showy undersides. The female is a drab brown, and therefore inconspicuous when laying her eggs on heathland vegetation.

tions has its own invertebrate wildlife, including a variety of spiders.

Walking on water

One of the most impressive heathland carnivores is the raft spider, which lives on and beside the small pools of water that occur where heath turns into bog. Lurking at the side of a pool, it uses the tips of its feet to test the surface tension of the water for vibrations, in the same way that other spiders use their finely-strung webs. As soon as it feels movement on the water, it races across the surface to the source of the disturbance. If an insect has fallen onto the water, the raft spider will quickly devour it.

Heathland spiders include both wolf spiders – which scuttle around on the ground finding their prey by sight – and the superbly camouflaged crab spider. The male crab spider is drab and small, but the female is colored bright pink, giving her perfect camouflage among the heather flowers as she waits for her prey. When an insect approaches, she stretches out her front legs. As soon as it passes, she lunges forward and grabs her prey.

The crab spider tackles prey much larger than itself, including bumble bees, and is sometimes lifted into the air if it has caught a particularly strong individual. The victim will not go far, however. The crab spider's poison is powerful, and the victim will die soon after being bitten. The two creatures then fall to the ground, and the crab spider sucks its prey dry.

Although they are powerful predators, spiders are not immune from attack from other animals. One of their main enemies is the solitary wasp, which provides its young with live food. Before laying its single egg, the solitary wasp digs a burrow in the soil and goes off to find prey, such as a spider or caterpillar. Once it has stung and paralyzed its victim, the solitary wasp carries it into the burrow, where it lays an egg on the still-living prey. The wasp then seals the burrow. When the wasp grub hatches, it feeds on the food its mother has provided.

Heathland specialist

One of the most colorful of heathland insects is the silver-studded blue butterfly which takes to the wing during the summer, when it may occur in large numbers. Unlike many invertebrates found on heathland, it is restricted to the heath habitat.

The female silver-studded blue lays her eggs on the stems of gorse and heather, so that the caterpillars

ABOVE **The meadow pipit is a common heathland bird, although it is not restricted to the habitat. It shares its physical adaptations – a thin bill and strong feet – with many of the several thousand species of perching birds that feed on insects. It is a resident in Europe, but a migrant over some of its range, and this may be why it has longer wings than many other perching birds.**

can later feed on their soft, tender shoots, and – in the case of gorse – on the flowers. The caterpillar produces a sugary substance known as honeydew from its body, and this attracts ants, which cluster around the caterpillar to feed. The ants take the caterpillar into their nest for the winter, where it pupates, protected from the attentions of predators. In summer, the adult butterfly emerges from its pupa, and leaves the ants' nest.

The green hairstreak butterfly is related to the silver-studded blue, and also occurs on heaths; its caterpillars, too, are attractive to ants. However, the green hairstreak is not restricted to heathland, and occurs wherever its foodplants, which include gorse and broom, are common.

Heathland moths

Of the many moth species that inhabit heathland, the most striking is the emperor moth, a relative of the Oriental silk moth. The female emperor moth is large and purple-gray, while the male is smaller and redder. Both sexes have huge, eye-like markings on their wings, evolved as a defense against predators. (Many moths and butterflies have developed these eye-shaped patterns for the purpose of confusing or startling a predator long enough to give the potential victim a chance to escape.) The large green and black caterpillars of the emperor moth feed on heather foliage.

The ground below the heather is not as well-populated as the ground in habitats with less acid and nutrient-poor soils. The invertebrates that do live there include millipedes, primitive insects such as springtails and bristletails, and earthworms. They are hunted by small mammals, such as shrews and voles, and small birds, such as the stonechat and meadow pipit.

Reptiles of the heath

The sun-warmed, sandy banks, and the gentle shade that heather provides, makes heaths an ideal habitat for reptiles. All of the six species of reptiles that occur in the British Isles are represented on lowland heaths: they are the smooth snake, the sand lizard, the viviparous lizard, the slow worm, the viper and the grass snake. Four of these reptiles live in other habitats as well, but the smooth snake and the sand lizard are confined largely to lowland heaths (see box on page 236).

Other sun-loving reptiles, such as green lizards and wall lizards, occur on heaths further south in Europe. The snakes of the small American heaths include the northern brown snake, which has often been seen in urban areas.

In contrast to the viviparous lizard, which gives birth to live young, the sand lizard lays eggs, which it buries in a sunny place in early summer. The warmth of the ground incubates the eggs, and they hatch in the fall. The sand lizard feeds on invertebrates, catching

Heathland Reptiles

Sand lizards and smooth snakes inhabit open woods, hedgerows and grassy places in many parts of Europe. The north-western limits of their range reach as far as the British Isles, where the reptiles live only on the lowland heaths of southern England.

Like the sand lizards and smooth snakes, many animal species find few suitable habitats as they move towards the edges of their ranges. The habitats themselves gradually change: woods, for example, become lusher further north. Animals from more southerly regions are less suited to such environments and cannot compete successfully with the woods' other inhabitants.

In southern England, and in other parts of north-west Europe, heathland is the prime habitat of the sand lizard. Most other northern habitats provide neither abundant food nor the warm, soft sand the lizard needs to incubate its eggs.

Vanishing habitats

Little is known about the smooth snake's relationship with its environment. In southern England, it reproduces with more difficulty than it does elsewhere in Europe, probably because of the cooler northern climate. Female smooth snakes in England breed only once every two or three years. Even when they do breed, the snakes may produce still-born young or experience difficulties shortly before birth. Further south, however, smooth snakes breed every year, and with greater success.

The sand lizard, the smooth snake, and other heathland animals such as the silver-studded blue butterfly, the red-backed shrike and the Dartford warbler, are at the edges of their ranges in north west Europe. If they are to survive, conservation measures will have to be stepped up to protect their vanishing heathland habitat.

BELOW Smooth snakes need to bask in the sun to raise their body temperatures high enough for them to become active. The open character of heathland provides many areas for basking. It is particularly important that pregnant females bask, since the warmth helps to incubate the developing young.

ABOVE The whinchat visits the northern European heathlands in summer to breed, but flies south in winter when the invertebrates on which it feeds are in short supply. By contrast, its close relative, the stonechat, which also breeds on heaths, moves to coastal habitats where the weather is milder and the invertebrates more abundant.

much of its prey – especially beetles and spiders – among mature heather bushes. The sand lizard is well-camouflaged as it clambers around among the heather's twigs and foliage. Even so, it often falls prey to the foxes and kestrels that hunt across the heath's open landscape.

The smooth snake – itself a rare reptile – preys on sand lizards, slow-worms (legless lizards) and small snakes. It either seizes its prey in its jaws and swallows it whole, or coils around the victim to immobilize it, before swallowing the prey.

Heathland birds

The three species of birds that nest on lowland heaths in north-western Europe are the Dartford warbler, the stone curlew and the nightjar. The Dartford warbler needs a habitat that supports both heather and gorse, since it builds its nest in the heather and searches for invertebrates to feed on in the gorse thickets. In spring, the bird emerges from the top of the gorse bushes from time to time to sing.

Unlike most other warblers that breed in the British Isles during the summer, Dartford warblers do not migrate south for the winter. They rely on insects throughout the year (rather than varying their diet with fruits and seeds in colder months), and consequently are especially vulnerable to severe winters when fewer insects are about. Then the number of Dartford warblers falls drastically. In Mediterranean areas, the bird lives on the coastal scrublands.

The stone curlew breeds on grassy heaths and in a range of other dry, grassy habitats. In recent years, it has also adapted to nesting on agricultural land. The bird has very large eyes, and is most active at dawn and dusk when its eerie, piercing call can be heard well into the night. The nightjar, in contrast, is truly nocturnal, pursuing insects over heath and bog at dusk and after dark. By day, it lies on the heathland floor, where its dull mottled coloration camouflages it against the brown vegetation.

Other common heathland birds include the skylark, green woodpecker and wren. The cuckoo often inhabits heaths, probably because of the abundance of the meadow pipit, a bird that frequently serves as a host species to the cuckoo by raising its chicks.

Dragonflies and hobbies

Dragonflies often hunt over heathland, catching and eating smaller insects. Several dragonfly species breed in ponds in damp heathlands, where their larvae eat other aquatic invertebrates. When the dragonfly larvae are about to metamorphose into adults, they climb up the vegetation beside the ponds and wait for their skins to split so that they will be free to emerge as adults.

Large hawker dragonflies fly over heaths in search of food, tilting and twisting in the air. As they do so, they may themselves fall victim to another aerial, heathland predator – the hobby. Hobbies are small falcons, about 1 ft long. In winter, they leave western Europe and migrate to Africa, returning in summer to hunt small birds, such as swallows and martins, as well as dragonflies. Hobbies often hunt at the edges of woods, especially those that back onto heaths. The nature of their prey reflects their flying ability; to catch a swallow on the wing demands great speed and maneuverability in the air.

Amphibious life

Heaths and moors are home to a variety of amphibians. The palmate newt occurs on lowland heaths and in moorland ponds. In continental Europe, the natterjack toad lives on heaths, although in the British Isles it is mainly restricted to sand dunes. Smooth newts and common toads breed on heaths, but occur in a wide range of other habitats too. The North American heaths have their own amphibians, including spadefoot toads. These creatures live in sandy soils, digging tunnels with their spade-like feet. They emerge at night to feed on invertebrates.

The restricted range of vegetation on heaths and moors is largely due to the lack of nutrients, especially nitrogen, in the soil. Insectivorous plants have adapted to this ecological challenge by using animal life as their source of nitrogen. Sundews are typical insect-catching plants. Their leaves are fringed with slender, red hairs that are tipped with drops of sticky fluid designed to trap insects that land on the plant. As the insect struggles to escape, the plant slowly closes its leaf around the insect. Digestive enzymes secreted by the plant extract all the insect's nutrients, and when there is nothing left of the plant's victim, the leaf opens again and the driedout insect – now a mere husk – blows away in the wind.

The beetle and the bug

Heather faces a virtual yearround attack from the heather beetle, which eats its leaves, and sometimes defoliates the plant so that it weakens and dies. The heather beetle lays its eggs in damp leaf litter or moss in spring, and the grubs feed on the heather shoots. The pupae develop in the late summer, and the beetles emerge in autumn. They continue to feed on the heather before hibernating for the winter. When they emerge in spring, they eat yet more of the heather before mating and laying their eggs.

The numbers of heather beetles are controlled to an extent by another heathland inhabitant, the pentatomid bug, which eats the heather beetle's grubs. Even so, in some years, the heather beetle may destroy entire stretches of heather moorland.

In the most of Europe, red deer live in woodlands, but in western Scotland, they graze on wet upland heaths where heather is less dominant and they can feed on grasses, sedges and rushes. Rabbits have become much less abundant on heaths since the advent of the viral disease, myxamatosis. As a result, trees have flourished on some heathlands where the young saplings have been left to grow, unharmed by the rabbits.

Fit for a grouse

On some moors, controlled burning is carried out in order to provide a habitat suitable for the red grouse, an important game bird. The bird thrives in areas that contain a mixture of old and young heath. The young heath provides it with the tender young shoots of heather on which it feeds, while the older heather is used for nesting and shelter. To create the right conditions, parts of a grouse moor are burned each year to ensure that there is a good mixture of old and new heath for the bird.

In winter, the red grouse feeds on virtually nothing but heather. During the rest of the year, it varies its diet with bilberry and crowberry leaves and fruit, as well as other plant matter.

Like the red grouse, the mountain hare feeds almost exclusively on

RIGHT The buzzard often hunts over moorland, but is also seen in the vicinity of woods and crags. On broad wings, it soars upwards on thermals (rising currents of warm air) to a great height, from where it can search the moorland for carrion or live prey. Like its relatives, the eagles, harriers and falcons, the buzzard has strong, sharp talons and a curved beak with which to tear at flesh.

heather, although if the weather is particularly hard and the heather is buried under snow, it will nibble at gorse shoots and willow bark. The mountain hare is always at risk from the two fiercest predators of the moorland, the stoat and the wild cat. Wild cats hunt from the cover of vegetation, crouching low and pouncing on their prey.

The red fox is another fierce moorland predator, while predatory moorland birds include the hen-harrier, the buzzard, the red-backed shrike, the merlin and the short-eared owl. The merlin is a fast, low-flier that captures small birds, while the hen-harrier, buzzard and short-eared owl all kill and eat hares and rabbits. They also kill grouse and their young, and this has resulted in their persecution by gamekeepers on grouse moors. Both the merlin and the hen-harrier nest on moors, but they spend the winter elsewhere, most often on coastal marshes, and sometimes on lowland heaths.

The red-backed shrike has declined as a result of climatic change and habitat loss. Widely known as the butcher bird because of its habit of sticking its prey onto the spines of bushes, such as hawthorn, it was once common on heaths and in scrubby places. Although always on the edge of its range in southern England, it is now rare throughout western Europe.

Heathland conservation

Ironically, as long as there is demand for grouse shooting, areas of moorland will be preserved. It is, at present, still a fairly common habitat. However, lowland heaths are one of the most threatened habitats in the world. Most of the heathland in north-western Europe has already been lost, although fragments, such as parts of Luneburg Heath in West Germany, have been saved as nature reserves. The constant threats of agricultural improvement for grazing, plowing for arable land, and urban development hang over all the remaining heathlands of Europe. Many of the heaths have been lost to conifer plantations, while others have become overrun by birch trees and lone pines. The pine seeds were originally carried onto the heaths from surrounding plantations, destroying the unique nature of the heathlands.

ABOVE In winter, the red deer develop longer and thicker coats to protect themselves from the bitter cold of the open moorland. At this time, the stags use their antlers to break off the branches of evergreen trees, such as holly, so that they can eat the foliage. During winter, male and female red deer often form separate herds.

Among the largest of all owls, the Eurasian eagle owl is superbly adapted to its role as a boreal forest predator. A thick layer of feathers offers the insulation it needs for nightly hunting forays even in the coldest winters.

BOREAL FORESTS

The stately fir, larch, spruce and pine trees that make up the vast boreal, or northern, coniferous forests, thrive in a cold climate and infertile soil. The forests harbor an array of animals, many of which lead solitary lives

LEFT About three inches long, the goldcrest is the smallest of the forest birds. It often nests at the end of branches where it is safe from tree-climbing predators.
PAGES 242–243 Stretching in a broad green band across the northern land masses, the coniferous forests are largely the product of a cold, continental climate. At first glance, much of the forest might seem empty of animal life, even during summer. Closer inspection reveals a surprising number of permanent residents and seasonal visitors.

A broad band of coniferous forest, known as boreal forest, stretches for 1200 mi across the upper region of the Northern hemisphere. Ranks of trees stretch towards the horizon in all directions. The trees grow so densely that they limit visibility in places to 40 ft or less. The boreal forests have evolved largely as a result of a cold, continental climate and they are absent from the Southern hemisphere, where there are no sizeable land masses at corresponding latitudes.

The penetrating cold of the long winters and the scarcity of heat and light have produced a habitat in which only specially adapted plants and animals can survive.

Another important ecological factor is water. In eastern Siberia, water is scarce because of low rainfall and because much of the precipitation is in the form of snow, which rapidly forms ice. However, in other areas, water is so plentiful that it prevents dead plant material from decomposing. The wet environment leads to an accumulation of such debris as peat, which eventually forms bogs.

In order to survive in such an inhospitable place, the plants and animals avoid competition with other species. Because of climatic conditions, the forest structure is more simple than in other habitats, such as the rain forest. These two facts together mean that only a narrow range of plant and animal species inhabit the forests.

The great forests of the north form the largest remaining wilderness habitat in the world, after Antarctica. Large areas are still virtually undisturbed by human settlement and they provide food and shelter for great numbers of animals.

A conifer forest offers its inhabitants a solitary existence. Food resources are scarce, and many species vigorously mark out and patrol their territory. The food-web (the collection of all the inter-related food-chains) is precariously balanced, and the ecosystem as a whole is extremely sensitive to population pressures.

Interrupted only by sea

Boreal forests are named after Boreas, the Greek god of the cold north wind. In Eurasia, they are widely known by their Russian name – taiga.

The forests stretch across the two main land masses of the Northern hemisphere – northern Eurasia and North America. They range from the edge of the tree-line in the north, where they merge into the virtually treeless tundra, south to the margins of the temperate woodlands and grasslands. The coniferous forests would form an evergreen belt right around the globe, were it not for the wide barrier of the Atlantic Ocean, and the narrow Bering Strait.

In North America, the boreal forest forms a broad curve that reaches from Alaska in the west,

right across Canada, to the northern part of New England in the United States. In central Canada, the icy influence of Hudson Bay pushes the tree-line southwards, causing a noticeable notch in the forest's distribution map. The taiga covers much of northern Scandinavia and European Russia, then widens into the denser taiga of Siberia and parts of China and Korea.

The boreal forest is a very young habitat. Within recent geological time (as little as 14,000 years ago) the area that is now occupied by the coniferous forests was covered by ice sheets and glaciers. Only when the glaciers retreated at the end of the last Ice Age, about 10,000 years ago, was the underlying land exposed for colonization by plants and animals. The process

ABOVE Conifers grow in dense stands, monopolizing the available light and leaving little for other plants. Mixed conifer stands are not uncommon; here, fir trees and larch share the available sunlight. The most nutritious parts of a conifer are the seeds, which are protected by a tough cone (inset). The spruce cones seen here have already opened and scattered their seeds onto the ground. Only animals with suitable adaptations can take the seeds from an unopened cone.

continues even today; in parts of Alaska, stands of spruce forest now grow in the wake of glaciers that have retreated within the last 200 years. The lack of diversity in species is attributed to the relative youth of the habitat, while the number of possible colonists has been limited to those species already adapted to cold winters.

Climate and seasons

The shape and form of the coniferous forests in high latitudes are influenced by several climatic factors. The most important of these is the cold. Air temperatures often fall below −30 F, and in parts of Siberia, temperatures frequently reach −50 F. They remain below freezing for much of the year and the mean annual temperature in this region can be as low as −5 F. Much of Siberia and parts of northern Canada and Alaska lie in the realm of permafrost. The ground is permanently frozen down to a considerable depth, and only the top few inches, known as the active layer, ever thaw.

Rainfall varies drastically with geography. The climate of Canada and Scandinavia is generally moist, with rainfall averaging between 15 and 30 in per year. In eastern Siberia, however, the yearly total averages between 6 and 12 in, and can be as low as two to 5 in, a rainfall figure that is comparable to those of many arid desert regions. Snowfall also varies considerably; parts of Canada receive up to 15 ft annually, whereas over much of eastern Siberia, snowfall rarely exceeds 8 in. In many areas, much of the water that could be available for use by plants and animals is locked into

the useless, solid form of ice. Thus, drought is one of the major factors that determines the character of the boreal forests.

Another factor is the unequal length of the seasons. Winter lasts for almost eight months, and summer offers only a brief respite. The length of day varies considerably and the boreal forest is sunk in perpetual gloom for at least half the year. During the brief summer, however, nights are short, and under almost constant sunlight, temperatures can rise to 80–100 F. Plants take maximum advantage of such a short growing season, and they grow mainly in the months between May and August.

Evolutionary fitness

Conifers possess a number of characteristics that have enabled them to become the dominant life form of the cold northern latitudes. The soils left behind by retreating glaciers are acidic and relatively infertile, and conifers have evolved shallow root systems that are a major advantage in colonizing such soils. They also have slender trunks and branches, which provide little resistance to the strong north winds that blow down from the polar regions. Their tapering shape enables them to shed snow easily before sufficient weight builds up to break their branches.

Long before the evolution of broad-leaved trees, conifers had developed the thin, waxy needles that are beneficial for several reasons. The needles have small surface areas that minimize heat loss and reduce transpiration (loss of water during leaf respiration), which is vitally important when there is little water available. When the temperature falls below freezing, sap also freezes, damaging the plant tissue. However, conifer needles contain only a small amount of sap, and because the needles are so thin, there is less plant tissue so frost damage is reduced to a minimum. For this reason, they can still photosynthesize when temperatures are only slightly below or above freezing, gaining maximum advantage from even the weakest sunlight. Only the larch, which grows in the coldest, driest conditions, sheds its needles annually so that it loses no water at all.

LEFT The micro-habitats created by areas of marsh and open water play an important role in the overall ecology of the coniferous forests. Here, near the Yenisey River in south-central Siberia, a tangled mass of deciduous vegetation grows on the water margins. During the short spring and summer, these plants and trees provide a valuable food source for many of the forest's herbivores, and the open water is home to countless insect larvae. During the winter, when the average temperature drops to −5 F, only the conifers remain green against a landscape of ice and snow.

Establishing dominance

Coniferous trees are well suited to the cold climate. They have established a stable dominance by altering a major physical component of the new habitat – the soil. The evergreen needles remain on the trees for up to seven years, and by the time they fall to the forest floor, nearly all the nutrients have been leached out of them. The deep carpet of dead needles, twigs and cones that is characteristic of boreal forest floors has virtually no food value. It decays slowly

LEFT Lichens can endure extremely low temperatures, and are common throughout the coniferous forests. In winter, they are an additional food source for deer and smaller animals. RIGHT Where conifers have been cleared by humans or fire, some deciduous trees – such as these silver birch – are able to establish themselves. They form an important element in the forest's food-web, because the leaves, shoots, bark and flowers (or catkins) are all edible. The green catkins (top inset) are female; the brown ones (bottom inset) are male.

and the cold further inhibits bacterial decomposition, making it difficult for other plants to grow. These conditions create the typical boreal forest soil type, known as podzol. The soil is characterized by high acidity and rigid layer formations.

Rain water percolates through the topsoil – a thin layer of black humus known as mor – eventually creating weak acids. These filter down through the soil, taking with them vital minerals and nutrients. The filtering process creates a layer of sterile sand that roots find difficult to penetrate. Above the sand, the topsoil tends to be water-logged. The soil layers remain stratified because the acidity of the soil and the intense cold inhibit the activities of earthworms and other soil mixing animals.

Even hardy conifers would find it difficult to sustain vigorous growth in such poor soils, were it not for a symbiotic relationship they have formed with certain species of fungi. The thread-like fungi that live among the litter on the forest floor are unable to photosynthesize, and must obtain their nutrients from living trees. By attaching themselves to the fringes of the conifers' root systems, they provide the trees with water and essential minerals in return for carbohydrates. Some estimates suggest that the trees' efficiency increases by up to 40 per cent as a result of this mutually advantageous relationship.

As far as the eye can see

Conifers can completely carpet whole landscapes if they are left undisturbed. Stands of a single species cover hundreds of square miles without interruption. Different species prefer slightly different soils; spruces, for example, thrive in heavier, damp clay soils, whereas pines and larches favor lighter, drier, sandy soils.

Conifers monopolize the available light to the detriment of other plants since their thick foliage creates dense shade. Spruce and fir trees create the clarkest shade, and this means that the vegetation on a spruce forest floor consists only of fungi and other plants that do not photosynthesize. Pine and larch forests are well-lit by comparison, and they tolerate a wider variety of small neighbors, but coniferous forests do not encourage dense undergrowth. The typical boreal forest floor vegetation consists of horsetails and sphagnum mosses in boggy areas, along with cowberries, bilberries, sedges, lichens and dwarf shrubs.

Plants that need light to thrive and grow tall are doomed to failure in boreal forests. Wood sorrel is one of the few flowering plants able to tolerate the deep shade. However, areas that contain few conifers play an important part in the overall boreal forest ecosystem. Sections of coniferous trees are periodically removed. In nature

this occurs either through the many flash fires to which the densely grouped trees are particularly prone, or through storm damage. During the last few hundred years, humans have created many more clearings through their need for softwood timber, which is obtained from conifers, and is used for building and furniture-making.

Exceptions to the rule

Hardy deciduous trees, such as aspen, willow, alder and birch, are the first to exploit these new clearings, their seeds being carried by the wind. Deciduous trees can achieve local domination for up to 150 years until the conifers, which eventually grow taller, reassert their domination. Such clearings, at varying levels of development, are scattered across the boreal forests. The broad-leaved trees, shrubs, grasses and other plants that grow in them form an important part of the forest's food resource.

The countless ponds, small lakes and acidic bogs that dot the landscape are important to the overall ecology of the boreal forest system and to the large areas of western Siberia that are covered by marsh. Although they represent separate micro-habitats, these areas encourage the growth of a considerable amount of non-coniferous vegetation. Aquatic plants, in particular, provide an important source of minerals for some of the region's herbivores. The widespread availability of open water during the summer months is also important for several species of the region's insects and birds.

The forest changes to a considerable degree on a north-south axis. In the broad middle band, the stands of trees are dense and tall. For example, spruce trees typically reach 50–60 ft with as many as 15 trees per 100 square yards. Towards the northern boundaries, the trees become progressively shorter and more widely spaced. The forest thins out almost imperceptibly at the band where the tundra and taiga overlap. Stands of trees, such as dwarf birch and larch, occur where local sheltered conditions permit, but they are separated by areas of bleak, windswept open ground. On the southern margins, where the conifer's dominance yields to the temperate mixed woodlands, the two habitats blend almost imperceptibly. Here, broad-leaved, deciduous trees penetrate deep into the dark, evergreen forests along sheltered valleys.

Overlapping habitats

The ecotones (transition zones between two plant communities, with some characteristics of each community) are home to much of the animal life found in the coniferous forests; few species occur exclusively in the boreal forests.

Ecotones often have a richer variety of organisms than either of

RIGHT Red deer are fairly common in the European areas of the taiga, and, like all deer, they can be destructive in their eating habits. During winter, when food is in short supply, the deer gnaw bark from the trees, exposing the white, living tissue (seen in the tree in the foreground). During summer, many animals, from bears to birds, feed on the succulent fruit of forest shrubs; these include the cranberry, bilberry and related bearberry whose pink, springtime flowers (inset, left) are succeeded by red berries (inset, right).

the bordering communities. The blurring between habitats is largely due to the scarcity of winter food, which forces animal species to range farther afield than their relatives in more hospitable climates.

Animals also move between the major habitats at different seasons. Animals from the north move southwards to find food and to shelter from the bitter Arctic winds. In summer, southern species move northwards to take advantage of the abundant food supplies produced during this short, but productive, season. Sometimes this movement takes the form of regular annual migrations, like those of the large herds of caribou and reindeer, or the invasion of insectivorous birds.

A few animals use the coniferous trees as shelter and as a hunting ground. Others are able to exploit the trees as primary food sources, some to the point of developing extreme specialization. However, most animals that live in the boreal forests have a much more wide-ranging diet. The other main animal adaptations are determined by the long, cold winter, rather than the forest vegetation, and none of the adaptations are exclusive to the boreal forest.

King of the forest

The brown bear (known as the grizzly bear in North America) is faced with only one enemy, namely man – an enemy that has severely reduced the bear's number and range. Placed at the top of the forest food chain, the larger species of brown bear stand almost ten feet tall on their hindlegs, and can weigh 1,600 lbs when fully grown. Brown bears prefer the deep woods that surround clearings, marshes and lakes. The open spaces provide most of their food, whereas the dense forest provides them with a refuge from man.

For the most part, brown bears are genuine omnivores and their main diet consists of berries, nuts, leaves, fungi, birds' eggs, insects, fishes and carrion. However, in the harsher northern regions of the forest, bears also hunt large deer.

In order to acquire sufficient food to sustain its huge size, a bear spends much of the year as a solitary creature, ranging over individual territories that cover hundreds of square miles. During the summer months, when prey and vegetation are abundant, it consumes extra food, building up reserves of fat to sustain it through the winter.

Brown bears spend the cold months in a den, which they either scoop out of the earth, or build among rocks and beneath fallen trees. Bears are not true hibernators, and although they spend winter in a kind of perpetual doze, they quickly awaken at any sound of danger. Brown bears mate during May and June in most areas, but the

ABOVE The red squirrel is the most common rodent in the coniferous forests, and during winter is able to survive by feeding on the seeds of cones that have not yet opened.
LEFT The pine marten is found throughout the taiga, and is the most agile of the Eurasian members of the weasel family. Although it usually builds its den on the ground, the pine marten captures most of its prey among the branches, and therefore does not need the camouflage of a white winter coat.

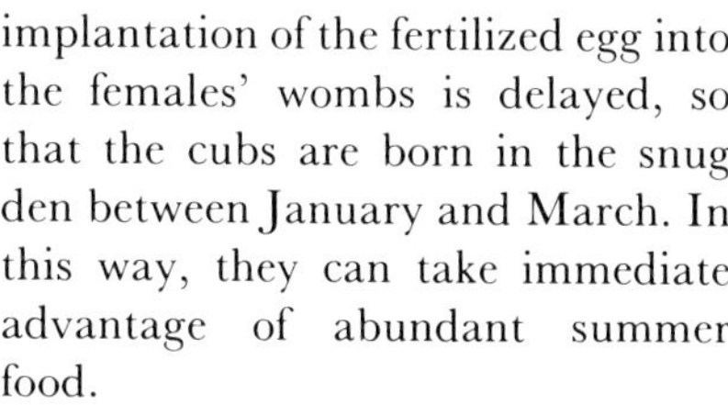

implantation of the fertilized egg into the females' wombs is delayed, so that the cubs are born in the snug den between January and March. In this way, they can take immediate advantage of abundant summer food.

The twig eater

Grazing land is scarce in the boreal forests, but despite this, the largest resident herbivore is also the largest member of the deer family. The Eurasian elk (known as the moose in North America) needs to consume 20 lb or more of vegetation every day in order to stay alive. Elk range widely throughout the summer. They live alone, preferring the clearings that are rich in grasses and young deciduous trees. They also frequent the marshes and river banks, where they feed on water plants. At the height of the summer, when vegetation is abundant, elk consume up to 40 or 50 lb of food every day in order to build up reserves for the cold months ahead.

In winter, the elks' rations are halved and consumption falls to as little as 15 pounds per day. Their winter diet consists largely of willow, aspen, and birch shoots, together with the young foliage of the pine. (It is this habit that caused the American Indians to name the animal "moose" – meaning "twig eater.") Elk also consume large quantities of bark, especially pine, spruce and aspen. Because of their feeding habits, elk can cause serious damage to the forest, and in Europe, during the fall they often damage oat fields.

Elk cannot walk in snow that is more than about 1 ft deep and they are often forced to move away from the more open areas which are potential sources. During these bleak wanderings, the elk try to avoid established trails wherever possible. All the animals are weakened by hunger, and the young and old become easy prey for wolves and other predators.

Basement level

The smallest mammals of the boreal forests spend their lives on, or beneath, the forest floor. Shrews, voles, wood lemmings, chipmunks and birch mice live in shallow tunnels, and emerge only to feed. During winter, voles shrews and lemmings remain active beneath the blanket of snow, feeding on insect larvae, roots and emergent shoots.

Chipmunks build up stores of food during the summer, and spend the winter months in their burrows, alternately dozing and feeding. Birch mice prefer to hibernate through the winter.

During hibernation, the animal switches off the mechanism that regulates its internal temperature, so that its body heat changes in relation to the temperature of the surrounding air. Although a sharp fall in temperature normally wakes a hibernating animal, the extreme conditions of the boreal forest mean that large numbers of hibernating animals freeze to death each year.

The common red squirrel exploits both the trees and the forest floor. In summer, red

A Fierce Forest Carnivore

The wolverine is the largest member of the weasel family, and occurs throughout the boreal forests and its edges. Combining size, strength and aggression, the wolverine has little to fear from any of the forest's other inhabitants. Even brown bears are wary of the damage a wolverine can inflict.

The wolverine's thick, shaggy winter coat is a good insulator and efficient at shedding snow; this enables the animal to burrow deep into snow-drifts to create its winter den. After mating in spring, the implantation of the fertilized egg is delayed so that the young are not born until the following spring.

Solitary and highly territorial, a single adult wolverine may range over 1,200 square miles of forest. Although agile, it rarely climbs trees, and takes most of its food on the ground. In summer, the wolverine becomes a cautious, nocturnal scavenger, since its size bars if from efficient hunting. Nevertheless, it does manage to supplement its diet of carrion with birds' eggs and small mammals.

The glutton

In winter, the wolverine is in its element: the animal's wide and thickly furred feet spread its weight, giving it an enormous advantage during a chase across snow. The wolverine can easily run small deer and foxes to ground, and may even drive a lynx or wolf from its rightful prey. In fact, the only animals safe from the wolverine are wolves and bears. Having made a kill, the wolverine feeds and then hides the carcass up a tree or in its den.

BELOW The wolverine's strength and fearlessness make it a formidable predator. It will attack an animal twice its size, such as a lame or young elk. It has even been known to drag a lynx from a tree by its tail. The wolverine is also called the glutton, referring to its habit of killing more than it can eat in times of plenty and storing the remains.

squirrels forage mainly on the ground for fallen seeds, although they also take birds' eggs and nestlings. In winter, they confine themselves to the trees where the unfallen cones – which present little problem to their chisel-like teeth – provide their main source of food.

Small mammals form the main food source for many of the forest's carnivores, such as the red fox. An extremely successful forest animal, it has more than a dozen subspecies widely distributed throughout the coniferous forest belt. All the other small mammal predators belong to the weasel family: boreal forest species include sable (found only in the USSR), marten, wolverine, stoat (or ermine) and the North American fisher. These predators are all solitary hunters. Some, like the fox and sable, take their prey of birds on the ground, or burrow beneath the snow for voles and shrews. Others, especially the pine marten, are agile tree climbers that stalk squirrels and nestlings among the branches.

The lynx and the hare

The widespread lynx is well adapted to life in the boreal forest where it feeds on a variety of animals. In Europe, for example, lynx feed on birds, roe deer and reindeer. The North American lynx also has varied feeding habits, but in winter it preys mainly on the snowshoe hare. During summer, when food becomes more abundant, the lynx is less likely to prey on the hare, but in winter, they often engage in one-to-one battles and surprisingly, the two are quite evenly matched.

ABOVE The lynx is a powerful predator of birds and small mammals. It often sits on low branches, waiting to drop on unsuspecting prey. The wide pads on the lynx's feet spread its weight and enable it to walk over snow without sinking too far in; only the snowshoe hare, which has similar adaptations, can outrun the lynx across deep snow. The small mammals that remain active beneath the snow are also at risk from the lynx. Using smell and hearing, the cat patiently trails its victim from above, before diving down through the snow to make a killing.

The lynx is a powerful and aggressive hunter. It lies in wait on a tree branch, then pounces on an

ABOVE The forest floor, especially the relatively open and grassy floor of pine forest, provides a home for many small mammals. Most numerous are the various species of vole, such as the northern red-backed vole. It spends most of its time in shallow tunnels, and emerges only to feed on leaves, shoots and seeds. During the winter months, the voles, along with the lemmings and the insectivorous shrews, remain active beneath the blanket of snow and feed daily in order to replace lost heat and energy.

unsuspecting hare. The hare, almost perfectly camouflaged in its winter coat, has wide, hairy pads on its feet that enable it to run quickly across the snow without sinking in. The feet of the lynx show similar adaptation, but its greater body weight slows the lynx down. Given any sort of head start, the hare usually wins.

Birds are much more successful than mammals at exploiting the resources of the coniferous forests, and they occur in a much greater variety of species. Their ability to fly, enables them to range wider and travel faster than the earthbound mammals. Birds can also exploit food sources that are out of reach of even the most agile four-legged climbers. The variety of species increases greatly when vast numbers of seasonal migrants arrive, taking advantage of the abundant food supplies of the forest summer.

The predatory bird species, such as owls and hawks, are at the top of their particular food-chains. The goshawk initiates its attack from a perch in dense cover and is capable of twisting and turning through the trees at high speed. It consumes a varied diet that changes throughout the course of the year, from place to place, and from year to year, depending on its availability. The goshawk's prey varies in size, from voles to hares, and it sometimes catches other birds of prey, such as sparrowhawks.

Several species of owl, including the great gray owl, the hawk owl and the North American boreal owl (known as Tergmalm's owl in Europe) live in the boreal forest and dominate it at night. The great gray owl protects its nest so fiercely that not even bears attempt to take the youngsters.

Perfect adaptation

Some widespread coniferous forest birds display a high degree of adaptation. Each of the four species of crossbill that occur in Europe feed off a different type of tree. As their name suggests, the birds' bills cross at the tips, enabling them to extract conifer seeds. Each species is

ABOVE **The capercaillie is the largest of the many gamebirds found in the coniferous forests of northern Eurasia, the male growing to almost a yard in length. It feeds mainly on buds and conifer needles. In winter, the western capercaillie digs a hole in the snow and stays there for most of the day, relatively safe from predators and sheltered from the icy winds. In eastern Siberia, where the snowfall is much less, the black-billed capercaillie must perch on branches, exposed to goshawks and other predators.**

adapted to feeding on a particular kind of cone: the large-billed parrot crossbill and Scottish crossbill are adapted for the large, hard, pine cones; the common crossbill for medium-sized spruce cones; and the white-winged crossbill for small, soft, larch cones.

The nutcracker removes cones from trees, takes them to the nearest rock or suitable branch and, placing one foot on the cone, hacks away at it with its beak until it frees the seeds. Like its relative, the jay, the nutcracker stores food for the winter. During summer, the nutcracker may spend all day collecting as many as 160 seeds in a pouch beneath its tongue. It then hides the seeds in caches in the ground. In winter, it demonstrates an amazing ability to find these stores, even under three feet of snow.

Few species of animals are as well adapted to life in trees as the woodpeckers. One of the largest in the world, the black woodpecker, occurs in the deepest, darkest parts of the taiga, and in mixed forests and parklands. Unlike most other woodpeckers which nest in dead trees, it often makes its nest in healthy trees. During winter, it occasionally supplements its normal diet of ants and wood-boring beetles and their larvae with conifer seeds and berries. In spring, it sometimes drills into trees to drink the sugary sap from beneath the bark.

Ground dwellers

Perhaps the most typical and widespread of the coniferous forest birds are the various members of the grouse family. The ruffed grouse and spruce grouse are found in North America, while the northern hazel grouse (hazel hen), the black grouse and the two species of capercaillies – the western and the black-billed – occur in Eurasia. During the summer, they roam widely, feeding on shoots, seeds and conifer buds; in winter, some species can subsist on pine needles.

The newly hatched young eat many insects during the first week or two of life, but they later become increasingly herbivorous, like their parents. To survive the cold winters, the forest birds usually burrow into the snow. The hazel hen digs a new hollow every day, and spends as long as 20 hours each day completely immobile in it to conserve energy.

In the taiga, the courtship rituals of the capercaillies take place each year on the same communal display grounds, known as leks. It is the only time that the males tolerate each other's presence. During elaborate displays in front of a group of admiring females, the males may be completely oblivious to anything else, and will even ignore shots fired by hunters.

In his courtship display, the male spruce grouse of North America fans his tail, erects the bright red "combs" (fleshy crests) above his eyes, and beats his wings rapidly. He may even utter a series of deep hoots to reinforce his visual display.

ABOVE The nutcracker is found only in the Eurasian taiga. Although its staple diet is pine and spruce seeds, it will also search the forest floor in summer for berries and insects. In winter, the nutcracker relies on caches of conifer seeds that it has hidden in the ground, beneath bark and in clumps of moss and lichen.

Cold-blooded inhabitants

Amphibians are not well represented in the coniferous forests, since the cold climate is an unsuitable habitat for animals that cannot adequately regulate their body temperatures or conserve heat. In the slightly warmer regions of the forests further south, however, the common frog and the common toad are widespread wherever water is available for spawning.

The only reptiles present in the boreal forests are the common viper and the viviparous lizard, but again, they are largely confined to the southern stretches. Both species give birth to live young, which is essential when there is not enough sunlight to warm the eggs.

Insect life

Insects are considerably more successful than amphibians at surviving in the forests, and occur in great numbers. During the spring and summer months, the air hums with a host of flying moths, beetles and wasps. Huge swarms of midges, blackflies and mosquitoes rise from the surface of rivers, ponds and marshes, hungry for the vertebrate blood they need in order to reproduce. Voracious caterpillars and larvae often cover trees, shrubs and plants, and one tree may harbor thoustands of caterpillars of a single species.

The inscets reproduce at a great rate, ensuring that enough breeders survive winter, and enabling them to withstand losses to predators.

PEST TO MAN AND BEAST

Human visitors are often surprised by (any suffer greatly from) the enormous numbers of biting insects, especially mosquitoes, that infest the summer forest. The mosquito has adapted remarkably well to the boreal habitat by developing ice-resistant eggs that can endure long months frozen into ponds and marshes. In summer, vast clouds of hungry adults emerge, often driving elk and other deer neck-deep into water to escape their tormenting bites.

In order to reproduce, the boreal mosquito species need to feed on vertebrate blood. But, for every cloud of insects that is fortunate enough to encounter a warm-blooded animal, such as a deer, countless other mosquitoes emerge into a landscape practically devoid of suitable animal life. As a consequence, mosquito populations are huge, giving them a greater chance of encountering something on which to feed.

No escape

The boreal mosquitoes have become highly sensitive to minute changes in the carbon dioxide levels produced by the breath of animals. Nevertheless, they will home in on the heat generated by their targets; studies have shown that cold-climate mosquitoes can detect differences in air temperature as little as one hundredth of a degree. If warm-blooded life is close by, the mosquito will find it.

A combination of numbers and efficiency has created in the mosquito a fearsome forest pest. In northern Canada, the attack rate of a large swarm can be as high as 9,000 bites per minute. For an unprotected human, this would result in the fatal loss of 50 per cent of total blood supply in less than two hours.

BELOW A single mosquito, such as the one seen here biting the back of a hand, may only cause a slight irritation to its victim. But in the northern forests, clouds of these ferocious biters make a daunting enemy that can force deer to take refuge almost antler-deep in water. Far from being inhibited by the north's extreme climate, the mosquitoes positively thrive there.

ABOVE **The ichneumon fly is, in fact, a parasitic wasp that usually lays its eggs within the body of a caterpillar or other living insect. Upon hatching, the wasp larvae immediately start to feed on their host, invariably killing it. Because the wasps parasitize a specific host species, some species have been used in experiments to find an ecologically safe method of controlling the insect pests that attack coniferous trees.**

Specialized feeding

Several insect species have adapted to life in the forest by feeding on the most abundant food resource – conifer needles. However, most insects have widely different strategies for surviving the winter.

The pine looper moth mates in midsummer, and the female lays her eggs on needles near the top of the tree. The eggs hatch after two weeks, and the caterpillars immediately begin to feed. As they move downwards, they can easily strip the whole tree of its needles before dropping to the forest floor in the fall to pupate. They emerge in the following spring.

Near the southern margins of the forest, the pine processionary caterpillar moth chooses to stay high in the trees during winter. The caterpillars have large heads with powerful jaws, ideal for grinding up pine needles. In October, the caterpillars form small groups at the ends of branches where they spin insulating silken cocoons in which to spend the winter. In spring, they move down to the soil to pupate, emerging in the summer in time to breed.

The sawfly has also adapted well to the boreal habitat, with different species feeding exclusively on pine, spruce and larch. The female lays her eggs at the tops of trees in late summer, where they remain until spring when the larvae hatch. They feed until midsummer, then descend to the soil where they spin themselves cocoons. Some emerge as adults within three weeks, while others suspend their development by entering a form of hibernation known as a diapause that lasts until the following spring.

Tunnels in the trees

Wood-eating beetles have successfully colonized the forest, and several species are widespread. The ambrosia beetle bores deep into trunks, where the larvae are able to feed off the black fungi that grows along the sides of their tunnels. The engraver beetle remains just below the bark, where it creates the familiar star-shaped pattern of laying chambers.

Wood wasps are widespread in the forests. They display a certain amount of specialization in their diet: the yellow-black species attacks Scots pine and spruce; the steel-blue variety favors larch and silver fir. The adults fly from June to the fall and after mating, the female uses her ovipositor to bore holes about one inch deep into the wood where she lays her eggs. When the larvae hatch, they bore deep into the tree; in small trees they may even reach the center. The larvae remain feeding in the tree for up to three years before they are ready to pupate.

Ichneumon flies, members of the same group of insects as wasps and bees, occur throughout the boreal forests. Their parasitic larvae use the vast numbers of caterpillars and larvae as hosts, feeding on and growing inside or on them. Ichneumon flies usually live off a specific host species, and at least one type makes exclusive use of wood wasp larvae. By a process that is still not fully understood, the female ichneumon fly can detect the exact location of the larvae beneath the bark. Piercing as much as an inch or so into the wood with her ovipositor, she then lays her eggs directly into the hapless larvae host.

Tiny scavengers

The warmer regions of the boreal forest form an ideal habitat for wood ants. The deep carpet of dried needles provides an endless supply of building material for their nests, which are often formed around an old tree stump. The nests can reach a height of 3 ft and cover several square yards.

Wood ants have evolved ingenious methods by which they combat the cold climate. The gently convex shape of the mound is designed to absorb as much as possible of the little sunlight that exists, as well as to conserve heat. Some of the heat is generated by the decomposing pine needles. So effective are the ants' efforts, that there can be a difference of up to 40 F between the temperature inside the nest and the temperature of the air outside. Scientists believe that much of the heat in the nest is actually generated by the metabolic processes of the ants themselves.

Wood ants obtain much of their food in the form of honeydew that collects in the abdomen of plantsucking aphids. Apart from milking aphids, wood ants are also wide ranging scavengers. They can travel along the highest and flimsiest of branches, and they are able to coordinate the transport of prey much larger than themselves. A single wood ant community of about 100,000 individuals has been observed to carry more than 300,000 prey insects each day into the nest.

Such scavenging activity has important consequences for the overall health of the forest, for the wood ants control the number of potential pests. In regions where there has been large scale defoliation (leaf loss) because of heavy infestation of moths and sawflies, the areas around wood ant nests stand out as small islands of greenery.

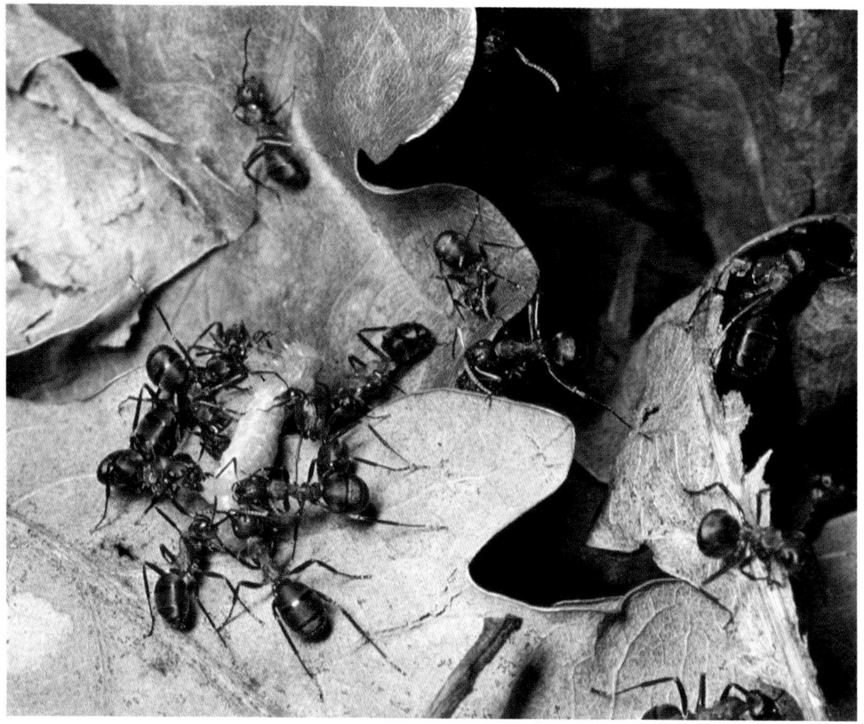

ABOVE Wood ants are probably the most useful of all forest insects, since their constant scavenging plays an important role in maintaining the health of the forest. Although it has not yet been proved, it is believed that wood ants are the only insects that use their body heat to maintain a steady temperature inside their nests. Exactly how cold-blooded insects manage this remains a mystery.

North American taiga

The boreal forests of the New World stretch from the southern coast of Alaska, across the Canadian shield, to Newfoundland and Labrador Island. Only in the vicinity of the Great Lakes and New England do the true boreal forests dip slightly into the United States. The coniferous forests of the western coast (characterized by giant Douglas firs and sequoias), extend as far southwards as northern California, but they are not considered true boreal forests. These lush woodlands benefit greatly from the moist maritime climate and are more properly termed temperate rain forests.

The pattern of vegetation is remarkably uniform throughout North America's boreal forests. The dominant coniferous species are interspersed with stands of paper birch, quaking aspen and balsam poplar, which occupy recently cleared ground. A distinct geographical variation in the coniferous species does occur. In the west, black spruce, alpine fir and lodgepole pine predominate, but as

LEFT Coniferous forests cover about 40 per cent of Canada's land area. The different species are remarkably successful at colonizing steep, rocky slopes. Here, the shallow lake has been fed by melt-water from the mountains, and the trunks of drowned trees can be seen on the "islands" in the middle. During the winter, the lake forms a frozen expanse of ice and snow, and is shunned by wildlife because it offers neither food nor shelter.

one moves eastwards, these are gradually replaced by white spruce, balsam fir, jack pine and the eastern larch. Nearer the tundra, dwarf birch and willow constitute the "land of little sticks."

The boreal forests cover nearly 1.7 million square miles of Canada, or about 40 per cent of the country's total land area. Innumerable ponds and stagnant pools, commonly known by their native Indian name, muskegs, are scattered throughout the forests. Varying in size from a few hundred square yards to 50 square miles, these water-logged areas form micro-habitats that are important to the overall ecology of the boreal forest. The water provides a habitat for the larvae of many forest insects, and allows for the permanent growth of deciduous trees and some flowering plants, such as Labrador tea.

Hefty herbivores

In North America, the non-migratory woodland caribou shares its boreal forest home with the moose (known in Eurasia as the elk). The caribou takes its name from an Indian word meaning shoveler, since the animal uses its large hooves to clear away winter snow as it seeks out the mosses and lichens lying beneath it. Wapiti (the American name for the red deer) also range into the southern margins of the boreal forest during the summer months. Although usually confined to the temperate mixed forests, wapiti often push northwards in the competition for food, to share in the forest's summer abundance.

Another occasional visitor is the white-tailed or Virginian deer, which occupies roughly the same ecological niche as the European roe deer. Although fundamentally a grazing animal of woodland edges and other habitats, white-tailed deer can cause considerable local damage in the forest by browsing on the shoots and saplings of deciduous trees.

The rarest inhabitant of the forest – and the largest terrestrial mammal in North America – is the wood bison, locally known incorrectly as the buffalo (it is no relation to the true buffalo). It was once thought to be extinct, surviving only in hybrid form as a cross between wood bison and plains bison. However, a herd of pure wood bison was discovered in 1957, sheltering in a remote corner of Wood Buffalo National Park, in Alberta, Canada. In summer, these immense beasts feed

RIGHT The North American moose (known in Eurasia as the elk) is a year-round forest inhabitant, usually solitary in its behavior. In summer, the moose prefers recently cleared areas where it can feed on grass and young shoots; it is also often found wading in open water and feeding on submerged aquatic plants.

on shoots, shrubs and tree bark, but during winter, they use their hooves and heads to clear away the snow from the scant ground cover of dried sedges and moss.

Porcupines and fishers

The North American porcupine can be found throughout the North American boreal forests right up to the Arctic tree-line. It is particularly well-adapted to life in the trees, using its strong claws to climb into the highest branches. It feeds mainly on the young wood directly under the bark, and favors young pine trees above all others. But the North American porcupine can subsist on herbs, mistletoe and even pine needles, giving the animal a range of alternative foods that enable it to survive throughout the cold winters when little food of any kind is available. In the Old World, the Eurasian porcupine has not colonized the taiga at all.

The porcupine has stout quills that ward off potential enemies. Even the fearsome grizzly bear is wary of it; a face full of barbed quills can stop it from feeding normally, reducing it to a state of starvation. Only one forest predator seems to have learned how to flip the porcupine onto its back, exposing the vulnerable underbelly – and that is the fisher, one of the largest of the martens.

Known to trapppers as the black cat, the fisher ranges right across Canada. It is the most agile of the tree-dwelling mammals and can outrun squirrels, snowshoe hares, and even other martens. The fisher behaves extremely aggressively, attacking and eating almost anything. For this reason, the fisher stands at the top of the tree-dwelling food chain.

Retreat to survival

The bald eagle is another predator that occupies the top of a food chain. Once widespread throughout North America, this magnificent bird is now largely confined to western Canada and Alaska. Ironically, the bald eagle has now largely disappeared from the rest of the United States where it is regarded as the national bird. In 1978 there were fewer than 4,000 bald eagles in the United States, compared with about 80,000 in Canada and Alaska. A century ago there were more than 250,000 bald eagles spread across the whole continent. The reasons for their disappearance include the destruction of their habitat, shooting and the effects of pesticides.

Thousands of bald eagles, retreating from human encroachment, have now adopted the boreal forests as their preferred habitat. The feed almost exclusively

ABOVE The American robin, which is actually a thrush, is one of the boreal forest's regular summer visitors. Wintering throughout the southern half of the United States, many robins fly north each spring to take advantage of the forest's brief glut of food. The breeding range of the robin covers the whole of the boreal forest, and extends over the southern parts of the tundra.

on carrion, but will take live prey if it is easy to catch. During the period from October to December, the bald eagles gather in large numbers in the trees overlooking the major salmon rivers, particularly the Chilkat River in south-eastern Alaska. Here, and on other rivers, they feed on the vast numbers of salmon that die after spawning. Salmon provide a valuable source of protein for other forest carnivores too, especially the grizzly bear.

The Eurasian taiga

Stretching for over 6,000 mi from the North Sea in the west to the Bering Strait in the east, the Eurasian taiga is the largest single forest on earth. In Siberia, the Yenisey River separates the taiga into two distinct regions based on the species of trees that grow in each.

In Scandinavia, European Russia and western Siberia – that is, west of the Yenisey – the dominant trees are Norway spruce, Siberian spruce, Siberian fir, and, on sandy soils, the pines. The most widespread pine species is the Siberian pine, commonly called the Siberian cedar. Its large seeds are a valuable food source, prized by both animals and humans.

East of the Yenisey River, the Daurian larch predominates, to the virtual exclusion of other species. The main reason is the extreme climate of eastern Siberia, which experiences the greatest annual temperature range on earth. In winter, the temperature plummets as low as −55 F whereas maximum summer temperatures of 105 F have been recorded. Throughout the year, most of the soil is hidden under permafrost, preventing most other trees from gaining nourishment. But the Daurian larch has evolved a very shallow root system that enables it to draw water from the thin, top layer of the permafrost which thaws in summer.

Wolves in the wild

One major difference between the wildlife of the Eurasian taiga and that of the New World forests is the existence of a far greater number of wolves. In North America, humans have largely driven the wolves out of the forests. For hundreds of years a campaign of extermination has been waged against the wolf, and the species is now under serious threat. In the Siberian wilderness, the wolf has

been extremely successful and there are at least eight subspecies.

Outside the breeding season, wolves live in large family groups or packs. Because the food requirements of the pack are large, wolves have become extremely efficient predators of large herbivores. Elk and other deer are their year-round prey, while the onset of winter brings a seasonal bonus of migrating reindeer.

During summer, the pack may range over hundreds of square miles without hindrance, but in winter they have to avoid soft, deep snow, and so they run their prey to ground over windswept, open ground. The wolf pack displays extraordinary cunning and coordination when hunting. One group may make a diversionary attack, scaring the prey and driving it towards an ambush; or the pack may

ABOVE The taiga extends over 6,000 mi from the Baltic Sea to the Sea of Okhotsk. Light penetrates these birch woods in Karelia (western USSR), but further east, the forest takes on the dark and gloomy aspect typical of the western taiga where spruce and fir trees predominate. East of the Yenisey River in central Siberia, larches – which are better suited to the drier eastern climate – take their place. Conifers such as the Scots pine and Siberian stone pine occur throughout the taiga.

ABOVE The sable once ranged right across the taiga, but human demand for its luxurious fur pushed it to the edge of extinction. Now found only in Siberia, the sable is beginning to recover its former populations in the wild, thanks to the commercial farming of caged sables. Like many forest predators, sables are strongly territorial and will only leave their usual hunting ground when faced with starvation for themselves or their young.

split, with each side attacking simultaneously in a pincer movement. When searching for prey among the migrating reindeer, wolves usually single out old, weak and sickly animals as easier targets. By removing the stragglers, the wolves contribute to the overall health of the herd; those reindeer that survive are likely to make the best use of the forest's meager winter resources.

Unique species

A few inhabitants of the Eurasian taiga do not occur in the New World. One of these is the European bison, which now inhabits areas of Russian forest near the Baltic coast. The species had become extinct in the wild by 1925, but has been reintroduced from captive stock.

The sable, whose thick winter fur is prized above all others, is an inhabitant of the Siberian taiga. Sables are close relatives of martens and climb trees with agility, but they take most of their food on the ground. They prefer areas of pine forest, where they can supplement their normal diet of rodents with pine seeds in winter, and with berries and insects in summer.

The shy musk deer lives throughout southern Siberia, east of the Yenisey River. A solitary herbivore, it seems to have colonized the taiga from its more usual mountain homelands. The musk deer feeds mainly in the early morning and evening, taking leaves, mosses and grass. In this way it escapes the attention of the large predators that hunt mainly by day.

The largest of the world's cats, the Siberian tiger, is a rare visitor to the south-eastern taiga, since it is normally confined to the warmer, mixed forest of the Primorsky territory, in the extreme south-east of the USSR. However, the tiger's great range (each tiger needs a hunting territory of about 200 square miles) occasionally brings it into the coniferous realm. Today, the Siberian tiger hovers on the point of extinction, with only about 200 individuals left in the wild.

Far-eastern animals

In the Soviet Far East, the taiga covers most of the 600 mi-long Kamchatka Peninsula as well as

ABOVE The taiga, especially in Siberia, is the last great stronghold of the wolf, and the gray wolf is the most widespread subspecies. Wolves have been so successful because they hunt as a pack; the "lone" wolf is largely a product of popular myth. Most individuals will submit to the pack leader, rather than risk a solitary existence.

Sakhalin Island to the south. The flora of both regions is generally typical of the eastern Siberian forests; however, the animal life is remarkably untypical. The differences may hold some clues to the processes of animal colonization.

The Kamchatka Peninsula has no elk, musk deer, chipmunks, lynxes, Siberian weasels or other species that are typical of the rest of the eastern taiga. The reason for this is that a natural barrier of open tundra covers the northern neck of the peninsula, preventing many species from crossing further down into the peninsula. Species to which the open ground presents no obstacle, such as the snow sheep, the long-tailed Siberian souslik and the black-capped marmot, have been able to enter this region. Probably because of the absence of other large predators, and the consequent availability of food, the brown bears on the peninsula have become the largest omnivores in Eurasia.

The wildlife community on Sakhalin island, just north of Japan, lacks elk, roe deer and the Siberian weasel, whereas other more typical taiga species, such as the lynx, musk deer, chipmunk and flying squirrel, are present. These have managed to cross the narrow Strait of Tatar that separates the island from the mainland. Only five miles wide at its narrowest point, it freezes regularly, allowing the animals access to the island with relative ease.

The human impact

The coniferous forests have played a distinctive part in agriculture, trade and commerce. In parts of Asia and America, the coniferous forests border some of the world's most productive farmland, and large areas have been cleared of trees. The pressure to cultivate is most evident in Siberia, where hot summers extend the northern limits of many arable crops.

The thick winter coats of the forest mammals have been prized by humans throughout history, and have drawn hunters into the deepest parts of the forests. Yet, despite the inhospitable conditions, these hunters usually managed to draw a subsistence diet from the forest's meager food resources. During the 18th and 19th centuries,

REINDEER IN HUMAN CARE

Twice a year, large herds of reindeer (in Eurasia) and caribou (in North America) migrate between their summer grazing on the tundra and their winter refuge in the forest. The round-trip can be as great as 1,000 mi and natural herds of over 100,000 individuals have been recorded. On the northern edges of the forest, in Lapland and parts of Siberia, reindeer are tied into an association with humans that borders on domestication.

All purpose animals

The reindeer have long provided Lapps and other northern peoples with many of their material needs – skins for shelter and clothing; meat, milk, butter and cheese; and antlers, bone and sinews, used for making utensils. They are also tireless pack animals. In return, humans provide selected herds with protection from the wolf packs that roam the migration routes. They give care and attention to young and weak animals, and sometimes stockpile fodder for them against particularly severe winters when the animals may not be able to forage enough for themselves. These herds fare better than their usescorted "wild" counterparts.

Lapp lifestyles

The traditional nomadic way of life, in which Lapp families followed the migrating herds, has passed. Today, only herders accompany the animals – and even this way of life is under threat. City lifestyles and more comfortable jobs are a constant temptation for young Lapps, and the herding tradition is slowly declining. When the relationship finally ends, the most prolonged of man's activities in the boreal habitat will leave almost no trace.

BELOW In Lapland, reindeer have been semi-domesticated since about the 5th century, providing the Lapps with much of their material needs. In summer, the reindeer range across the bleak Arctic tundra, but in winter, they stay close to the conifer forests where they can find spongy lichen – known as reindeer moss – by digging through the deep snow.

large human populations moved eastward and westward into the hearts of North America and Eurasia. Animals such as bears, lynx, sable, ermine and some of the foxes provided large or luxurious furs, and so were hunted and trapped in great numbers. The prevailing fashion among the military for furs intensified the commercial demand for them.

By the turn of the 20th century, many animal species had virtually disappeared from Scandinavia and European Russia, and from large areas of North America. Fortunately, cheaper synthetic substitutes and the introduction of commercial fur farms have done much to reduce the threat to the wildlife. Conservationist legislation and public awareness, too, have assisted in their survival, and natural populations have regained their former levels in much of the forests.

A hunger for wood

Combining lightness with strength, coniferous softwood makes an ideal building material for houses, boats, rafters, joists and floorboards. The furniture industry also uses softwoods extensively to make such items as pine tables and chairs, as well as chipboard and blockboard.

In previous centuries, the use of traditional and less efficient logging methods, combined with quick-growing conifers, meant that there was little threat to the forests. But the widespread introduction of mechanization into forestry during the 20th century coincided with an explosion in world demand for softwoods, especially in the form of wood-pulp to make paper. Now, a single issue of a modern Sunday newspaper, with all its many supplements, requires approximately 150 acres of medium-sized trees. With this level of demand, the edges of the natural forest have quickly become denuded, and commercial forestry (in which trees are raised like any other crop) is now a major industry. While some commercial foresters create plantations, others continue to exploit the natural forest.

Harmful pesticides

Like all farmers, commercial foresters are keen to achieve the maximum yield from their land. Conifers, however, are prone to heavy and destructive infestation by species of moth and sawfly, and the foresters have used large quantities of pesticides against them. The results have been detrimental to all surrounding wildlife. A more natural method, and one that has been successful in certain areas, has been the introduction of moth and sawfly predators to the worst affected areas. But both methods of control affect the delicate natural balance of the ecology; the large concentrations of caterpillars and larvae are an essential part of the ecosystem, much of which depends on the summer food glut. Studies have shown that commercially

ABOVE The Siberian chipmunk is the most widespread of the Eurasian ground squirrels that inhabit the taiga. When foraging during spring and summer, the chipmunk, like the hamster, carries food in its cheek pouches. It takes some of this food back to its underground den to store for winter. During the cold months, the chipmunk remains in its burrow and spends much of the time sleeping, waking periodically to feed.

managed forests contain only approximately a third of the number of birds found in the natural forests.

Fuels and metal wealth

The forests of Alaska, Canada and the Soviet Union cover some of the world's largest untapped mineral reserves, especially oil. Exploration and exploitation of the most remote areas now take place in the face of strong conservationist opposition. Nevertheless, the industrial lobby is extremely powerful, and the exploitation of this hidden wealth increases year by year. The Alaskan and Siberian oil and gas fields, for example, are now in full production.

Although pipeline construction causes temporary physical damage to wide swathes of the forest, the damage is not permanent, and the ecosystem soon reasserts itself. In the past, pipelines were normally constructed running along the ground. However, in the case of the Trans-Alaskan pipeline, an ecological battle has been won; about half the pipeline has been elevated, averting the danger of the crude oil (which has an average temperature of 160 F) thawing the permafrost.

Lead and zinc mines

In the Yukon province of northern Canada, huge, open-cast mines have torn holes in the landscape, and the resulting mineral-laden dust has affected thousands of acres of forest. While conservationists protest, the authorities point to the fact that mining lead, copper and zinc accounts for 40 per cent of the province's economy.

Acid rain

One of the greatest threats to the coniferous forest – acid rain – comes from the skies, and is invisible. In the industrial nations, the large-scale burning of fossil fuels (coal and oil) produces huge amounts of sulfur dioxide; this combines with oxygen in the atmosphere to form sulfur trioxide, which in turn combines with water to form dilute sulfuric acid. It is this that forms the acid of acid rain.

Exporting pollution

When it rains, the acid has an immediate effect on the trees, often denuding them of foliage. More importantly, it drains into the soil, increasing its overall acidity. Countries cannot control what happens to the acid in the atmosphere because of the weather patterns; as a result, acid rain created in the United States and Western Europe falls mostly on the forests of Canada and Scandinavia, respectively. Since the chief producers of pollution largely export its harmful effects elsewhere, the problem has now become one of diplomacy as much as ecology. Despite considerable discussion, a comprehensive international agreement has yet to be reached.

Killing the birds

The continuing use of DDT as an insecticide to control crop pests has had serious global consequences. Because animals do not expel DDT from their bodies, it accumulates in their fatty tissues, becoming progressively more concentrated as it moves up the food chain. The effects of DDT on mammals are as yet little understood, but it has been definitely linked to declining populations among birds of prey, including the bald eagle. DDT also causes eggs to develop thinner shells, thus decreasing the chances of a successful hatching.

ABOVE The coniferous forests are highly valued as a source of fuel to combat the bitter cold of winter. Gathering fallen branches for firewood is a harmless activity, especially as the numbers of traditional forest-dwellers, such as the Lapps, are steadily declining. It is when humans begin to exploit and export the forests' reserves of oil and gas, that permanent damage to the habitat can occur.

Refuges and national parks

Although many of the problems threatening the coniferous forests are beyond the control of individual governments, large areas of forest around the world have been designated as national parks and wildlife sanctuaries. The governments of the United States, Canada, the Scandinavian countries and the USSR have all acted responsibly – not only preserving individual species and habitats, but also ensuring that future generations will be able to enjoy some of the most fascinating and beautiful wilderness places on the planet.

ABOVE Before the advent of mechanization, humans made relatively little impact on the forests. The small-scale use of the forest's resources were not harmful to the environment – Siberian lumberjacks, for example, used traditional methods and materials to build their cottages (above). However, modern machinery and chemical pesticides have vastly increased our ability to exploit the huge reserves of softwood. Only recently have people begun to appreciate that the forests are not limitless.

TUNDRA AND POLAR REGIONS

While winter in the polar regions conjures landscapes beset by blizzards and icy dangers, the brief spring and summer – especially in the treeless, subpolar tundra – bring an explosion of life to the wilderness

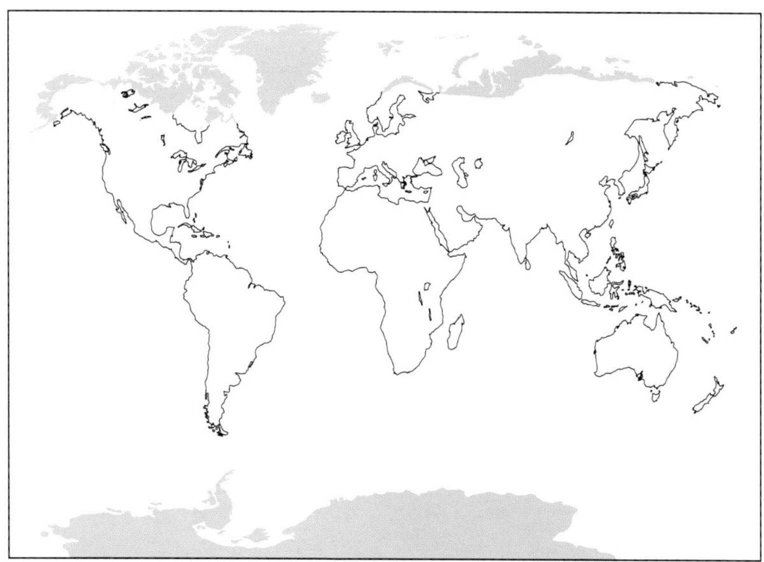

Contrary to popular belief, the Arctic and Antarctic polar regions do not start at the Arctic and Antarctic Circles for all purposes. The Circles simply act as geographical boundaries but they have no biological or ecological significance. Geographers find them especially useful for making comparisons between the poles: they often contrast the two regions, in order to highlight differences. The Arctic Circle rings an area that includes forests, farms, pastures, towns and villages, and a settled human population of at least two million people. In contrast, the Antarctic Circle encloses a cold desert continent devoid of trees or shrubs; it has no industries, no permanent settlements, and a transient human population that is numbered in hundreds.

Geographers drew the Arctic and Antarctic Circles 66 degrees 32' from the Equator, relating them to the angle of tilt of the earth in its annual passage around the sun. When ecologists need to define polar regions that have some biological meaning, they use boundaries other than the polar circles. Their Arctic starts at the treeline, the boundary between boreal forest and tundra. The Southern hemisphere has neither austral forest (the Southern hemisphere's equivalent of the northern boreal forests) nor tundra, and no treeline worth considering. Consequently, the ecologist's Antarctic region starts at the Antarctic Convergence – a line in the Southern Ocean that is often visible from the air, and which marks the furthest spread of cold Antarctic surface waters. The line appears where the colder, less saline water from the Antarctic meets the warmer, salty waters from the north. Both the treeline and the convergence are real, easily-identified boundaries that separate communities of plants and animals.

Polar living conditions

Within both the Arctic and Antarctic Circles, conditions become cooler and drier the closer one moves towards the poles. Climatic differences between seasons grow more extreme: summer days and winter nights become longer; summers are cool while winters grow very much colder; and when the sun disappears altogether – as it does every winter beyond the polar

ABOVE Thin snow covers the high Arctic landscape of Baffin Island for nine or ten months each year. The landscape is relatively new, having emerged from under thick ice sheets within recent geological time, some 10,000 years ago.

RIGHT The white winter coat of the Arctic fox is thicker and longer than its brown summer coat. Its superb insulating ability allows the animal to withstand temperatures of −60 F

PAGES 272–273 In the icy landscape of the polar desert, the sun hangs low in a clear sky, casting hard shadows and highlighting the surface of glaciers and snowfields. Beyond the Arctic and Antarctic Circles – the limits of the polar regions – the midsummer sun remains above the horizon day and night, but in midwinter the sun disappears, causing perpetual night.

circles – twilight or darkness reigns for weeks on end. In the dark season, photosynthesis and capture of energy by plants ceases altogether, since plants receive no light from the sun. At the poles themselves, the seasons and days merge as six-month days alternate with six-month nights.

Thick ice

Flowing water decreases as one moves towards the poles. Throughout much of the Arctic and Antarctic, water stands as solid ice for months on end, providing no benefit to plants or animals. For much of the year, thick ice stills the rivers and streams, while semipermanent pack ice, many feet thick, covers huge expanses of ocean. The ice limits the depth to which sunlight can penetrate and inhibits exchanges of vital gases.

The polar conditions impose similarity on two regions that would otherwise be very different: the Arctic region is a ring of fragmented, mainly low-lying land that surrounds four deep ocean basins; the Antarctic is a broad ring of oceanic water, complete but strewn with scattered island groups, surrounding a continent. An observer at the North Pole stands on an ice floe at sea level; one at the

RIGHT Arctic tundra soils, thin and rock-strewn, support only a meager vegetation of low shrubs, grasses, small flowering plants, mosses and lichens. On a corner of Iceland's coast, shown here, snow covers the land for most of the year. The vegetation is green in spring, but turns red-brown in the fall. Coastal plains are deserted in winter, but in summer they are nesting grounds for thousands of migrant swans, geese, ducks, waders and smaller birds.

South Pole stands 9,000 ft above sea level, on a continent-sized ice cap almost 10,000 ft thick.

Isolation and adaptation

Continental Antarctica and its fringe of islands have long been isolated from other continents by deep oceans. In contrast, Arctic lands have always been joined to temperate lands, and this has strongly influenced the number and variety of species that occupy them today. Northern plants and animals have always been able to cross and re-cross from temperate lands to feed in the Arctic. Plenty have failed, for Arctic cold and dryness eliminate all but the strongest. However, enough succeeded to establish the Arctic flora and fauna we now recognize. The typical flora of the region includes the various plant communities of the tundra and polar desert. The fauna includes caribou, bears, voles, lemmings, wolves, butterflies and spiders that forage among them.

Antarctica and the Southern Ocean islands are relatively empty, since few terrestrial plants or animals from southern temperate lands have crossed the oceans to colonize them. Although some of the Southern Ocean islands offer milder and more hospitable conditions than the Arctic, they remain poor in species. Temperate species of plants and animals introduced by humans, purposely or accidentally, have often been successful. The colder southern islands and Antarctica itself are even more isolated and empty.

There are many examples of the northern wealth of species and the southern wildlife deficiency. In summer, the northern tundra is ablaze with the colors of hundreds of thousands of flowering plants. Continental Antarctica, on the other hand, has only two species of flowering plants – a grass and a tiny pink plant – and the warmer southern islands support only a handful more between them. The flora of Devon Island, a small Canadian Arctic island, consists of 182 species of lichens, 132 species of mosses and 90 species of flowering plants. In contrast, the very much larger area of the maritime Antarctic (the Antarctic Peninsula and its neighboring islands) supports 150 species of lichens, 75 species of mosses, and just two drab species of flowering plants.

Contrasting wildlife

Arctic plants, soils, ponds and streams support a wide variety of invertebrates from more than a dozen major groups, including butterflies, bees, beetles, hoverflies and mosquitoes. Antarctica, however, has relatively poor soils and very little fresh water; its land fauna consists of a few species of micro-organisms, insects and mites. The continent's largest land animal is a wingless, mosquito-like fly.

The northern tundra is the year-round habitat of eight species of land birds and the summer home of more than 150 species that migrate in from the south. Antarctica has no land birds, the Antarctic islands have only a few resident species, and neither has the capacity to support migrants, except those that feed at sea. The Arctic has more than 40 species of grazing, browsing and predatory land mammals, ranging in size from voles to musk oxen.

While tropical deserts and rain forests may have occupied their present positions for tens of millions of years, polar ecosystems have had less than five million years to develop – a very short time span in evolutionary terms. Ten thousand years ago, during the last glacial period, many present-day tundra regions lay deep under ice sheets, while the forest edge lay far to the south of its present position. The tundra and polar desert we see today have existed only since the ice sheets retreated, although polar plants and animals began their evolution in an earlier period.

Polar climates

Latitude for latitude, the Antarctic is far colder than the Arctic, and its cold spreads farther towards the equator. The mean annual temperature at the North Pole is probably about −5 F – nobody has stayed there long enough to measure it accurately. At the South Pole, where records have been kept for more than 30 years, the mean temperature is close to −30 F. The high plateau of Antarctica is the coldest place on earth, and its chilling effects are felt across much of the Southern hemisphere.

Scientists at the Soviet ice cap station Vostok, 7,500 mi from the South Pole, recorded a temperature of −100 F – the lowest known temperature in Antarctica; Vostok takes the world record for low temperatures each year in late August. The lowest recorded Arctic temperature is −63 F – at Verkhoyansk, a small town close to the Arctic Circle in central Siberia (where similar lows occur each January). Both northern Siberia and central northern Canada are far colder in winter than the polar basin itself; Siberia is colder even than the Greenland ice-cap, which is very much higher in altitude. Paradoxically, the same Siberian and Canadian regions are among the warmest in summer.

Summer and winter alike, air temperatures between the Antarctic Circle and the South Pole are generally 30–40 F lower than those of equivalent latitudes in the north; low temperature is one of the main reasons for the relative emptiness of the region inside the Antarctic Circle. In warmer southern latitudes, differences with equivalent northern latitudes are smaller, but still striking. James Cook, the 18th-century navigator, was among the first to record such differences. South Georgia, which Cook rediscovered and charted, stands at 54 degrees south; his homeland in northern England stands at a similar latitude in the north. While Britain is a forested land with a mild temperate climate, Cook found South Georgia to be heavily glaciated down to sea level. On the whole island, he noted that "... not a tree was to be seen, nor a shrub even big enough to make a toothpick."

Simple ecosystems

The ecosystems of the Arctic and Antarctica evolved separately, since they have always been separated by the Earth's warm regions. Although the two ecosystems evolved in isolation, they have certain aspects in common. Both regions are young, immature and unstable (that is, they are liable to wide fluctuations in population numbers). Their ecosystems are generally simple, involving few species and uncomplicated food webs.

Polar lands cover wide latitudinal ranges and support many different plant communities. They are grouped according to their plant life, under the two main headings of tundra and polar desert. Each main group can be subdivided into two, and all four subdivisions merge into each other.

Polar oceans

The polar regions began to chill during the final disruption of the Gondwana supercontinent, when Antarctica split from Australia and drifted far from temperate latitudes and into its present polar position. (Gondwana comprised Antarctica, Australia and parts of South America, Africa and India.) The presence of high land at the pole generated the ice cap that slowly spread over all of Antarctica.

Both polar oceans formed long before the poles became cold. Seventy million years ago, the Arctic Ocean was a northward extension of the Atlantic Ocean, fringed by tropical trees and containing a fauna that included turtles and crocodiles. Ten million years ago, it had become a basin; beavers, forest horses and other animals browsed in the cool temperate woodlands that grew along its shores. The Southern Ocean was still subtropical twenty million years ago, when the first penguins found it. Its cool, fertile and hospitable seas attracted all kinds of creatures from neighboring temperate waters.

The two polar ocean systems differ in shape and structure. The Arctic Ocean, in its four deep, steepsided basins, is strongly layered and static; a layer of thick, irregular pack ice covers most of its center for much of the year. Although the ice drifts with winds and tides and is constantly renewed, it effectively cuts off more than two-thirds of the ocean surface from winds and gas exchanges with the atmosphere.

The Arctic Ocean basins contain little life; even in summer, explorers see very few seabirds and seals. In the area around the basins, sea ice comes and goes seasonally and winds and currents stir nutrients up from the depths, providing food for an abundance of marine wildlife. The surface waters support floating planktonic plants and invertebrates – notably copepods and euphausiid shrimps. These tiny creatures are the main food of fishes, inshore seals and seabirds. The sub-Arctic seas off Iceland and south-west Greenland, towards Svalbard, around North Cape, and off Alaska, east Siberia and Kamchatka, have long been known to humans as good hunting grounds for whales and seals. Their waters – which include some of the world's richest fishing and shrimping grounds – today support huge commercial fisheries.

The southern ocean

The southern ocean is more dynamic than the Arctic: it constantly exchanges waters with other oceans and is continuously stirred by winds and currents. The Weddell and Ross seas, and many of the lesser seas close to Antarctica, are biologically poor, since they have an almost permanent covering of pack ice. Other inshore seas, with seasonal pack ice, are richer in summer. Seals, penguins and petrels breed on the continent and on the inshore ice, forming large colonies close to where the water is richest and the food most readily available.

Ice-free waters

Away from land, the southern ocean is relatively ice-free. Surface waters in the belt of latitudes between 35 degrees south and 55 degrees south are driven constantly eastwards by westerly winds. Where their flow is diverted or interrupted – for example, east of the tip of Antarctic Peninsula and downstream from the islands of the Scotia Arc – nutrients well up to the surface from deeper layers. Such activity creates enormous eddies and patches of rich water where plankton proliferates – squids, fishes, birds, seals and whales congregate in the areas to feed on the plankton. Island groups of the southern ocean have a similar effect on local water circulation. Seals and seabirds breed on their beaches and cliffs, and feed in the neighboring waters. Some of the world's largest flocks of seabirds – mostly smaller relatives of the albatross, called petrels – feed from the seasonal riches of the southern ocean. Both the southern fur-sealing industry of the 19th century and Antarctic whaling of the 20th century drew their wealth from the southern islands and the seas close by.

ABOVE In the fall, polar seas begin to freeze over. By early winter, they are covered in a sheet of ice that extends far from the coast. Although often broken by storms or tides, the ice reforms and thickens with fallen snow, so that it measures five to ten feet thick by late winter. In spring, the sheet breaks into large, drifting floes, such as those shown here along the coast of the Chukchi Sea. The summer sun melts the floes' surfaces, creating pools of freshwater. Seals breed on the floes, sea birds rest on them, and algae, crustaceans and small fishes flourish underneath them.

Tundra

Tundra is the vegetation that is most typical of Arctic regions. The name, probably of Finnish origin, means "treeless heights." Originally describing the bare plains of the Kola Peninsula of northern Scandinavia, it was later used to encompass all the lands north of the boreal forest in Eurasia, North America and Greenland. The name tundra now applies to both the northern plains and to the vegetation that covers them.

Arctic tundra is a low-lying vegetation that is dominated by small shrubs, grasses and other flowering plants, with a sprinkling of mosses, liverworts and lichens. The vegetation of alpine tundra is similar to that of the Arctic tundra, and occurs in various forms close to the snow-line on high mountains throughout the world.

ABOVE The Weddell seal was named after James Weddell, a 19th-century sealer and explorer. These silver-gray seals occur on the sea ice close to the coast of Antarctica, and grow to 10 feet long. Though fearsome-looking when roused, they are normally quiet, peaceable creatures that like to bask in the sun. They are clumsy on land, but surprisingly agile in water. Weddell seals feed on fish and squid; in shallow, open water they hunt by sight, but in deep water, and in the dark under the sea ice, they use echolocation to find their prey. A layer of blubber or fat up to 4 in thick keeps them warm in icy seas. Sub-polar fur seals have a denser pelt that can be made up into fine furs. Many thousands of fur seals are killed each year for their skins.

Polar desert (again the name applies to both the location and the vegetation) has a much-reduced flora that occupies poor soils and receives little protection from winter snow. It is a tundra-like vegetation that lacks upstanding plants. Only the most hardy, low-lying species of flowering plants, mosses and lichens can survive in the dry, cold conditions.

The treeline

The treeline, where both Arctic and tundra begin, is a well-marked ecological boundary that weaves across northern Eurasia and North America. Tundra lies in a belt of varying width across Greenland, Iceland, Svalbard, the island groups of Franz Josef Land, Novaya Zemlya, Severnaya Zemlya, Novosibirsk and Wrangel in the Arctic Ocean, and the northern coastlands of Siberia, Alaska and Canada.

Along the northern coast of Scandinavia, the tundra forms a narrow band only a few miles wide between forest and shore. Along the Siberian, Alaskan and Canadian coasts, it broadens to about 400 mi wide. In Greenland, a narrow strip of rich tundra lines the southwestern seaboard between coast and ice cap, and a much poorer version appears beyond the ice cap on the northern coast.

Trees that can grow to heights of 25–40 ft in the shelter of the forest are stunted at the forest edge where they meet the tundra, twisting and crouching away from the wind. In some places, the treeline is narrow and sharply-defined: it is possible to stand among trees on the edge of natural woodland and see, just a few yards away, the start of the shrubs that indicate tundra. The trees also

end abruptly at the edges of streams or rock outcrops, on the brows of hills or at the start of a steep incline.

Hardy spruce

Some trees, such as spruce, are particularly hardy and can withstand the harsh conditions of the tundra. Consequently, their natural boundary extends much farther north. They form a broader treeline that often measures several miles wide. The boundary region contains a patchwork of stunted, ill-formed trees that gradually give way to open tundra. The uneven ground, marshy patches and permafrost cause the trees to grow unevenly. Islands of stunted spruce and pine alternate with shrubs, grasses, mosses and other tundra plants. In winter, deep snow protects the trunks, but leaves their growing points exposed to wind and frost-bite.

A shifting line

From one decade to the next, the treeline fluctuates, shifting north or south as a result of long-term changes in climate. The location of the treeline at any given time depends on the condition of the tundra beyond it. Although the trees at the edge of the forest produce seeds that germinate, the seedlings are not usually successful in the poor soil conditions of the tundra. When the conditions for regeneration deteriorate, the trees that die naturally at the forest edge are not replaced, and the treeline retreats south, enabling the tundra to spread. When conditions improve, successful seedlings advance from the forest edge into the tundra and the forest spreads for a time.

Bleak and bare

The single most striking characteristic of the tundra is its lack of trees. Much of it occupies ground that, within the last few thousand years, has been thoroughly bulldozed by ice-sheets. The relief is low, with rolling hills and few dramatic breaks. Few plants grow higher than a human's waist, and many lie flat along the ground. All try their best to evade the strong, biting, snow-carrying winds that blow unhindered across the tundra's immense plains.

For a few weeks, at the height of the brief summer, the tundra is ablaze with flowers, and is thronged by breeding birds and mammals and buzzing insects. Most of the birds are migrants that make long annual journeys to and from temperate or tropical wintering grounds. Many large mammals are also migrants, moving from forest to tundra, or from southern to northern tundra, to take advantage of the brief flush of seasonal growth during spring and summer.

ABOVE There are no trees or shrubs over the entire Antarctic continent; the densest vegetation stands only a few inches tall. Lichens, algae and mosses (seen here) are the most typical forms of plant life. Wind, cold and dryness restrict their growth, and they appear mainly in sheltered corners, where water from melting snow moistens the soil for a few weeks in the summer. Communities of tiny insects and mites live among the moss stems.

Soils and vegetation

The tundra close to the treeline is frozen solid in winter and often water-logged in spring and summer. Even in relatively dry areas, the winter snows accumulate. When the snow melts, the water forms fresh-water ponds and streams that fill depressions in the undulating surface. The melt-water cannot drain away naturally, since the subsoil remains frozen solid throughout the year, forming an impenetrable permafrost layer. Consequently, the tundra is often boggy, and many of the plant inhabitants are water-tolerant species.

ABOVE Arctic tundra vegetation is a patchwork of mosses, lichens and flowering plants, struggling to grow in thin, stony soils that often dry out in summer. Reindeer, musk oxen, hares, voles, lemmings and insects feed on these pastures, but tundra plants grow very slowly, and cannot support heavy grazing. Plants only a few inches high may be several years old.

In drier, well-drained areas, and in places where winter snowfall is slight, there is less standing water in summer and more bare, dry ground. In these areas, drought-tolerant plants predominate, sometimes growing in strange, circular patterns. The patterns are formed as a result of movements in the soil caused by repeated cycles of alternate freezing and thawing of the ground. These actions also lift the soil and grade its particles according to size, raising stones to the surface and limiting plant colonization.

Tundra soils consist mainly of rock dust and fragments with little organic content, and they are readily dispersed by the wind. Mosses and lichens that use the soils for anchorage are generally immature. Where flowering plants gain a footing, their roots penetrate the soil, bind it together, and release organic acids. Helped by bacteria, protozoans, nematodes, earthworms and other soil organisms, the acids convert the poor soil into a richer substance. Seasonal warmth and moisture encourage the process, but soil development normally takes many hundreds of years. The creation of soil is more advanced on the southern rather than the northern tundra; in the far north, soil development has barely begun.

Sub-Arctic tundra

The deepest soils, the widest selection of plant species and the highest densities of plants are found on the southern sub-Arctic tundra. The vegetation, like the soil, is irregular and patchy. The best areas support 3 ft high thickets of shrubs. Alder, birch and willow alternate with lush meadows of moss, grasses, heathers and more spectacular and colorful flowering plants. In early summer, the sub-Arctic tundra flowers – dwarf Arctic lupins, buttercups, windflowers, gentians, poppies, willow-herbs, chickweeds, vetches, campions, saxifrages and Labrador tea – carpet the land in a dramatic display. Many of them are familiar to gardeners as weeds or alpine plants. From July onwards, grasses and the leaves of shrubs turn red, while the bilberries, crowberries, bearberries and windberries that flourish on the tundra add autumn color to the scene.

On the sub-Arctic tundra, damp hollows alternate with drier ground to produce acidic bog soils that give rise to sedges, rushes, cotton grass, mares tails and moss mires. The mosquitoes and biting flies that breed in the ponds and streams, ferociously attack mammals, birds and people (if they are present). These insects spend most of their lives as eggs or aquatic larvae in fresh water, feeding on microscopic algae. As adults, only the females bite and a meal of blood enables them to lay a great number of eggs.

The richer tundra habitats at first appear to have solid carpets of vegetation similar to temperate grasslands – green in spring and brown or red-brown in the fall.

ABOVE In late summer, the dried-out bed of a lake in Iceland seems to be devoid of life. But the mud contains insect eggs and algae spores that will develop during the spring flooding. The winter's snow that accumulates on the hills melts during the spring thaw, and runs down to collect in natural basins such as this. The water also brings down nutrient-rich sediments for the abundant life that often occurs in these seasonal lakes.

However, closer examination reveals the carpets to be threadbare. Even the richest sub-Arctic tundra has a thin, incomplete covering of vegetation through which underlying rocks and bare soils emerge.

Arctic tundra

Away from the treeline, the tundra spreads for many acres seemingly without change. Moving north, the vegetation becomes thinner, and is dominated by dwarf shrubs, compact ferns, flowering plants pressed close to the ground, and carpets of lichens and mosses. There is no visible, sharply-defined boundary between the sub-Arctic and Arctic tundra, but the 45 F summer isotherm (the line linking places with the same temperature) offers a useful indication as to where one grades into the other.

In a more northerly Arctic tundra zone, most of the vegetation rises little more than 8 in from the soil. The main plants are dwarf willow, heathers and two or three species of avens, interspersed with short grasses, rushes, tightly-packed cushion plants and tiny heathers. Many of these plants appear fragile, ready to be blown away by the next storm, but they will have survived for decades or even centuries. Annual growth adds only a tiny amount to stems and subterranean roots, rhizomes and bulbs. A dwarf willow that is over 50 years old, for example, may have stems no thicker than a pencil.

Close to the ground

Only the hardiest plants can survive in the colder, drier conditions of the northern tundra. Although there is little winter snow, the winds are strong and the ground

ABOVE LEFT Cold, strong winds and dryness cause tundra plants to adopt special shapes for survival. Many, for example stonecrops and saxifrages, have rosettes of leaves that form a protective covering for the delicate growing points. Leaves are often tough-skinned, and covered with fine hairs to prevent water loss.
LEFT Many mosses and small flowering plants adopt the shape of the cushion plant. The body of the cushion is formed by many years of growth. From it spring the new shoots, packed tightly together for mutual protection against the wind. In spring, temperatures within the cushions may be several degrees higher than outside.
ABOVE Retreating ice sheets leave a gently undulating landscape of hard, bare rock, with hollows that form lakes and ponds. Ice-covered in winter, they thaw in summer and attract migrant ducks, geese, swans and divers.

is often warmer than the air above it. Most species of vegetation are small and compact. Cushion plants absorb heat both from the sun and from the ground. In direct or hazy sunlight, the temperature within a cushion plant is often several degrees higher than the air temperature outside, and stems, leaves and growing shoots benefit from the extra warmth.

At sub-zero temperatures in the northern winter, wind-blown snow becomes hard and crystalline – almost as abrasive as blown sand. Because the snow lies thinner on the ground in the most northerly region of the tundra, tall plants that rise above the snow level are severely blasted by the hard, driven snow. Only ground-hugging plants find protection from frost and wind in the thin snow. In the milder south, winter snow lies thick, and plants up to 3 feet high may benefit from its protection during the seven or eight hardest months.

Polar desert

Moving northwards into still colder, drier conditions, the tundra vegetation continues to thin out until it reaches polar desert. There is no sharp boundary between the two regions, but tundra vegetation that gives less than about 50 per cent cover is often called fellfield. Tundra that gives less than 20 per cent cover is called polar desert. The 35F summer isotherm shows roughly where the separation occurs. Only the most meager vegetation grows where mean temperatures for the warmest month stay below 35F.

Polar desert vegetation is typical of the coldest mainland shores, and of islands far out among the semi-permanent pack ice of the Arctic Ocean. The southern strip still contains some Arctic tundra, but in the northernmost desert, only the toughest species remain to form scattered mats among the wind-polished rocks and shingle. Lichens are the most prominent plants, especially the leathery, encrusting species that hold fast to rocks and even penetrate them. Among them grow mosses and algae, especially dark mats of blue-green algae that take nitrogen from the atmosphere and, in dying, make it available as a valuable nutrient to other plants.

Mosses, algae and liverworts grow in places that receive occasional water, such as snow-melt channels and patches of sandy soil that may be damp for only a few days each spring. These plants all need warmth and water from time to time, but they can withstand long spells of drought. When dried out, they are practically immune to frost damage. Flowering plants are still present, but only in tiny clusters: poppies, moss-campions, saxifrages and tufted grasses add a tiny dash of color to the summer landscape. Most of the color of this zone, however, is provided by the brilliant reds, oranges, greens and blacks of encrusting lichens.

Tundra birds

The Arctic in winter has little to offer land birds and only eight species are year-round residents. In spring, Lapland buntings, which winter in temperate and sub-Arctic areas, spread over the southern tundra. Snow buntings, which winter in the temperate regions of northern Europe (including Britain) and on the southern tundra, move north to breed. They eat mainly seeds in winter, but add insects to their diet in spring and summer. Common redpolls and Arctic redpolls are also seed-eaters, relying heavily on birch seed in winter. Rock ptarmigan and willow ptarmigan both winter on the tundra, digging in the snow with feathered feet for insects, seeds and young shoots. Their white winter plumage and brown summer coloration give them superb camouflage throughout the year. Ravens and snowy owls are predators and scavengers that scour the tundra all year for insects, birds and small mammals.

Breeding perils

The resident tundra birds breed in May and June, and rear their chicks through the summer in days of 24-hour sunlight. Within polar ecosystems, there are only a few species at each level of the food chain, while the range of food available is extremely limited. Unusual weather patterns can cause havoc. A late thaw, a late snowfall or an untimely cold spell may isolate whole species from their normal food supplies, starving many adults and chicks, and perhaps even destroying a whole season's breeding efforts.

Snowy owl populations fluctuate from year to year, according to the amount of food available for feeding the chicks. In poor years, few or no chicks may be raised over large areas. In years when lemmings and voles are plentiful, adult snowy owls may successfully rear clutches of seven or eight chicks. Young owls increase dramatically in numbers and disperse widely, appearing far south in temperate regions and in the northernmost polar deserts.

In summer, both the sub-Arctic and Arctic tundra bristle with birds. From April onwards, the migrants pour in from the south. More than 120 species take up residence all over the tundra, even in desert areas, exploiting the summer abundance of food. Swans, geese, ducks and waders (shorebirds) cross to Greenland from wintering grounds in north-western Europe; lesser golden plovers fly from Hawaii to Alaska and from Australasia to Siberia. Many migrants take the long-distance routes that cross the continents from south to north; gulls, terns, divers, sea ducks and other water birds move shorter distances from temperate zone to tundra, and from the coast towards inland regions.

A busy time

The migrant birds have much to accomplish in the short summer months, and they waste little time. They have usually fattened themselves before leaving their wintering grounds, and have already paired with a mate. Many return to familiar grounds where they have bred successfully before. They

RIGHT The snowy owls are the best-insulated of all birds. Found throughout the Arctic tundra, they patrol on silent wings – even in the far north in the depths of winter. They feed on voles, lemmings and other small mammals. In good summers, when the weather is mild and food plentiful, the snowy owls lay clutches of a dozen or more eggs, and raise large broods. In hard years, survival is difficult even for adults, and they may not breed at all.

ABOVE Early in the northern spring, golden plovers leave their wintering grounds in South America, Asia and the southern USA and head for the Arctic. Many make long flights over the sea, while others follow coastal routes. Three species breed on the northern tundra, and they arrive as the snow melts. In spite of their brilliant breeding plumage, the feather-patterns of the incubating birds blend well with the short vegetation.

often refurbish an earlier nest, and their first eggs are ready for laying within hours of arrival.

The foods that attract birds to the Arctic include buds, freshwater algae, the young shoots of grasses and aquatic plants, insect larvae and other invertebrates of soil, ponds and streams, beetles, crane flies, moths and fishes.

Among migrant predators are three species of skuas, gulls and several birds of prey. They feed on smaller birds, eggs, nestlings and ground mammals such as voles and lemmings, all of which become available in great numbers during the three summer months.

The availability of food depends mainly on the timing of the spring

thaw, when melting snow uncovers vegetation and ice disappears from the streams, ponds and lakes. The schedule is extremely tight, and bad weather or a late thaw can prove disastrous, especially for some of the larger migrant birds such as swans and geese. Even a short delay to the start of breeding may prolong nesting into the late summer, and seriously endanger the chicks' chances of survival. Long spells of bad weather that destroy eggs, nests and nestlings, or prevent breeding altogether, are fairly common. By July and early August, the juveniles have left the tundra region, and the parents undergo a complete molt before returning to their wintering grounds.

ABOVE Black-throated divers are one of four species of divers (loons) that winter on southern coasts and breed in the Arctic. In spring, they settle at the edges of large lakes where they spend the three summer months raising their chicks.

Tundra mammals

Of the 50 or more species of land mammals that live in the Arctic, about a dozen are common year-round residents of the tundra. These are all indigenous species that evolved from temperate stocks, and are now well-adapted for polar life. Ranging from musk oxen and caribou to tiny shrews, their stratagems for dealing with the tundra and its hazards vary according to their size and ways of life. The large mammals live out on the tundra all year round; the smallest disappear below the surface in winter,

MUSK OXEN OF THE ARCTIC

Musk oxen inhabit the Arctic and sub-Arctic regions of the far Northern hemisphere, where they live peaceably in herds of a dozen or more individuals. They take their name from the prominent glands below their eyes, which secrete a musky scent. When disturbed, the animals rub their scent glands against their front legs.

Musk oxen once grazed throughout the northern tundra, but after many years of human hunting activity (for meat), only small populations remain. Today, natural populations of the animals occur only in northern Greenland and some islands of the Canadian archipelago. Fortunately, musk oxen adapt well to human management and domestication. Efforts to bring back stocks of these fine animals to Siberia, Svalbard, Alaska and northern Norway have proved very successful.

A male stands about five feet tall at the shoulder and weighs up to 700 lbs. Its long fur, matted beneath the body and over the shoulders, is waterproof, windproof and provides efficient insulation against the cold; its stocky body and short limbs and tail further minimize heat loss.

Summer ease and winter hardship

In summer, musk oxen graze continually over wide areas of tundra. In winter, they seek ground where the snow cover is thin, and dig their hooves and muzzles into the snow, grazing on what food they can find beneath. Towards the end of winter, musk oxen survive mainly on reserves of fat left over from summer feeding. In extremely clod weather, they huddle together, sometimes under a vapor cloud of their own making.

In northern Canada, calves are born between April and early June. Through June and July, the adults feed voraciously, shedding their underwool and building up reserves of fat beneath the skin. During the rutting season – which occurs from July to early October – bulls, cows and calves move together in small herds. A few dominant bulls establish their status by bellowing and occasionally fighting with rivals. Wolves are the only serious predators of musk oxen. The herds protect themselves by forming defensive rings, placing the calves in the middle while the adults face outwards with lowered horns.

BELOW Musk oxen are the largest and heaviest tundra grazers, weighing as much as half a ton. Since they cannot escape the Arctic blizzards, they must withstand them as best they can. Their huge manes fill with snow, and their thick, woolly coats trail like ragged blankets, but the musk oxen stand four-square to the wind, facing the worst that the weather can produce. These creatures have only two enemies – wolves, and man.

and remain there until the warm weather returns.

Both mammals and birds maintain a high and constant body temperature, usually between 98 F and 102 F. Because of this constancy, these animals are known as homeotherms (the word means "same heat"). The heat needed to raise their temperature is generated as a by-product of the chemical activities that take place in the animals' muscles and other organs. The heat is stored in the body tissues and the energy used to maintain it is taken from food.

In cold, windy climates, homeotherms require a lot of energy to keep themselves warm because they constantly lose heat from their skin and through breathing. Polar homeotherms are well-insulated – birds are protected by their feathers, mammals by fur, and both store fat under their skin. Musk oxen and caribou, for example, have extremely dense fur on top of and beneath their bodies, and a thick layer of fat beneath their skin, which they accumulate during summer grazing. The fat also acts as a reserve of fuel to be drawn on during winter.

Huddling for heat

In the depth of winter, the fur and fat of large tundra mammals adequately protects them from the cold, but their insulation makes them ponderous and ungainly. They can maintain their body temperature in all but the coldest weather; in extreme cold, they huddle together to reduce heat loss. But their insulation is so efficient that, if they exert themselves by running, they soon overheat, and have to cool down by standing and panting. In so doing, they may find themselves at risk from predatory wolves.

Smaller mammals cannot afford such bulky insulation. Instead of confronting the winter cold, they evade it by living under the snow. Whatever the weather on the tundra above, at the meeting point between the snow and the ground, they are protected from the winds and intense cold. The dried grasses and seeds from the previous summer's crop provide abundant food, and they are relatively safe from enemies.

ABOVE Seagoing cousins of the brown bears, polar bears are creamy white, except for their noses. According to legend, they hide their black noses with one paw when they are hunting for seals on the sea ice, so as to be invisible. Polar bears are solitary creatures, roaming the tundra in summer and the inshore sea ice in winter, taking seals, birds' eggs, berries, fishes and carrion.

No Arctic bird or mammal hibernates, since hibernation would leave them vulnerable to the extremely low temperatures and to hungry predators. All the animals remain active, except for female polar bears, which enter dens and become torpid when they are ready to produce cubs. However, their body temperatures remain high, and they will awaken at the slightest disturbance.

Reindeer and caribou

The reindeer of Eurasia and the caribou of North America are different stocks of the same deer species. Several subspecies and local races, particularly among reindeer, are now almost entirely domesticated and bred selectively. Reindeer seldom grow to more than 3 feet tall, although large caribou bucks stand five feet high at the shoulder and weigh about 500 lb. The does of both reindeer and caribou are smaller than the males; both sexes have antlers. The animals usually have thick pelts for insulation against the cold, especially in winter. Several other species of deer inhabit the boreal forest, although they seldom appear on the tundra; the only exception is the elk, which often emerges to feed along tundra lakes near the treeline.

Caribou winter in small herds in the forest or on the southern tundra. Although some remain in the forest throughout the year, browsing on trees, grass and lichens, most form migratory herds of several thousand animals, which move north across the tundra in April. They feed voraciously on the spring growths of grass and lichens, growing new antlers as they travel. Calves born *en route* learn to walk and run with the herds from their first day. Wolves often accompany the caribou on their migrations and prey on the older and weaker members of the herd. The march north continues until August, when the caribou turn and start to move south again. Mating occurs on the move in September, and the herd is back at the southern end of its range by the onset of winter. Apart from the wolves, humans are constant hunters of caribou; both Indians and Inuit (Eskimos) hunt the deer for their fur, flesh, antlers, sinews and bones.

Moving with the herds

Reindeer, which followed similar patterns in Eurasia, were joined by human herders, rather than human hunters. In a curious symbiotic relationship, part protective, part predatory, herdsmen of many northern cultures – from the Saami in Scandinavia to the Eveni in Siberia – have, century after century, taken part in the migrations, controlling the movements of the herds and managing them in different ways. Some reindeer continue to migrate each spring and fall, while others are settled in compounds and herded locally like cattle. North Americans have never tried to herd caribou in the same way, although, from time to time, humans have introduced reindeer from the Old World in attempts to establish a pastoral industry. It has usually proved necessary to bring in Old World herdsmen too, for herding does not seem to come naturally to those whose tradition is hunting. Several large herds of reindeer have now been established in Canada and Alaska, to the benefit of local populations.

Other tundra mammals

The largest predatory mammals of the forest and southern tundra are the Kodiak bears – found only on a few Alaskan islands – and the grizzly, or brown, bears. Despite their fearsome (and by no means false) reputations as hunters of other mammals, full-grown adult bears are usually too cumbersome to chase prey. Instead, most tundra and forest bears settle for a quieter life of fishing, gathering berries, snuffling the ground for birds, eggs and offal, and digging for roots. Their behavior patterns closely resemble those of polar bears.

Gray wolves (also known as timber wolves) are the major predators of the tundra. They inhabit many parts of the world's far north, from forest to polar desert, and live in small communal groups. Whenever possible, the wolves prey on reindeer, caribou and musk oxen, usually singling out weak or sick animals. They also catch birds, dig small mammals from their burrows and feed on carrion.

In late spring, wolverines, coyotes, otters, lynxes, porcupines, red foxes and even muskrats leave the

RIGHT In North America, snowshoe hares live in boreal forests and along the edge of the tundra. Brown in summer and white in winter, they are well-camouflaged throughout the year. Their relatives, the Arctic hares, are larger and longer-legged. They have white coats all year round (with the exception of those living on the southern tundra which darken in summer). Arctic hares live in groups of up to a few dozen, usually where the snow is thin, so that they can forage even in the depths of winter.

forests to seek food on the tundra; the wolverines are especially adventurous, and may wander far north into the polar desert.

Arctic foxes scavenge and hunt throughout the year. In winter, their coats change from brown to white so as to blend in with the snow. Short-tailed weasels change color in the same way, their rich brown summer coats becoming pure white (with black-tipped ermine tails) in winter. Snowshoe, Alaskan and Arctic hares are the smallest of the mammals that live exposed on the tundra. Snowshoe hares, so-called for their massive feet, live mainly at the forest edge, while the other varieties – all closely related and possibly different races of the same species – range widely over the tundra. Some of the hares return to the forest in winter; others spend summer and winter in the far north, where food, however meager, is never far below the surface of the thin snow.

Shrews and lemmings

Zoologists have recorded eight species of shrews in the Arctic, mostly in warmer areas where insects are plentiful. While nine species of voles and three species of lemmings are listed for the Arctic, only the lemmings and two of the vole species inhabit the colder regions. The inhabitants of the cold regions are mid-sized rodents that grow larger than mice but smaller than rats. They have insatiable appetites for grasses, seeds and dwarf shrubs.

Lemmings have a high reproductive capacity and may breed for most of the year, producing four or five litters annually, each with five or six young. After two or three years of such breeding, they undergo a population explosion: the ground seethes with the small rodents in what has come to be known as a "lemming year." Skuas, owls, foxes and other predators move in from far and wide to take advantage of the easy pickings. Eventually, the lemmings eat all their local food supply and disperse. Although a few migrate successfully to other areas, most die of starvation, stress or disease. Lemming populations fluctuate locally in cycles of about four years – the overgrazed vegetation takes time to recover, and thus delays the start of the next cycle. Larger mammals, such as hares, show similar fluctuations of population for similar reasons; however, their breeding is slower, and their cycles extend over 10–13 years.

ABOVE Six species of lemmings inhabit the Arctic tundra. Resembling guinea-pigs and measuring 4–6 in long, they feed on grasses, seeds and the stems of shrubs and berries. In summer, they hide among the undergrowth, and in winter, they shelter in tunnels under the snow. Lemmings reproduce at an incredible rate; in favorable years, a single pair may produce twenty or more offspring.

Southern ocean islands

Continuous stretches of tundra and conifer forests occur only in the Northern hemisphere. The Southern hemisphere has no direct equivalents: in high southern latitudes where tundra and forest might be expected to occur, there is not enough land to support them, since the continents give way to ocean. Nearer the South Pole, only a few scattered islands in the

Polar Foxes

Two kinds of foxes live on the tundra. In winter, red foxes inhabit the temperate woodlands and the fields of Eurasia and North America. In spring, however, they migrate northwards as the snow disperses, moving towards the rich hunting grounds of the southern tundra, where they prey mainly on nesting birds. The red fox's rich brown coat, sharp ears tipped with black, and dark, slender brush-tail distinguish it from the Arctic fox.

The Arctic fox is slightly smaller and stockier than the red; it has round ears, silky fur and a great bushy tail. Unlike its temperate cousin, the Arctic fox ranges across many Arctic regions, and remains in the Arctic throughout both winter and summer. In spring and summer, the fox has a dull, gray-brown coat that blends in well with its background. After the fall molt, however, it takes on a much paler hue: Canadian Arctic foxes turn creamy white, while those of Greenland and Asia become blue-gray. Sadly, the hunting activities of humans threaten the continued existence of Arctic foxes. While both Inuit (Eskimos) and Indians continually set traplines for the foxes, the animals' furs, and especially their winter pelts, fetch high prices in First World fur-markets.

RIGHT Brown or blue-gray in summer and creamy-white in winter, the Arctic fox is protected from the wind and clod by its dense, soft fur. In the coldest weather, it curls up in the snow to keep warm. In both seasons, the animals patrol the tundra and inshore sea ice, sniffing out nesting birds and small mammals, and scavenging where wolves and polar bears have made a kill. Arctic foxes breed once or twice a year, raising three or four cubs when food is plentiful.

Masters of Arctic survival

Compared with the red foxes, Arctic foxes are the true masters of survival in the Arctic environment. Although too small to attack anything larger than hares, they are highly skilled hunters of birds and rodents. Throughout summer, Arctic foxes wander over the tundra alone or in pairs, digging for mice and lemmings, hunting young hares or snuffling on the ground for birds' nests.

Arctic foxes store their surplus food in caches for winter use. When hunting is poor, they follow polar bears and wolves, remaining at safe distances, but moving in at the first chance to scavenge on scraps. In winter, the foxes dig deep into the snow for lemmings and old carrion. Under favorable conditions, the Arctic fox may raise more than a dozen pups – it usually breeds in April and may produce a second litter before the end of August.

southern oceans provide a footing for vegetation.

The Antarctic Convergence, a marine boundary that can sometimes be seen on the sea surface, indicates the limit of spread of cold surface waters from Antarctica. Its presence affects land communities, since small islands derive their air temperatures mainly from the surface waters surrounding them.

South Georgia and the Falkland Islands, for example, lie in similar latitudes, but on opposite sides of the Antarctic Convergence. South Georgia, a periantarctic island (literally "around the Antarctic") washed by Antarctic waters, is very much the colder of the two. The Falkland Islands have a moorland vegetation of ferns and grasses, not unlike the rolling grasslands of neighboring South America. Their mature soils and more than 40 species of flowering plants have supported farming activity for several generations. The colder and more isolated island of South Georgia supports only 16 species of flowering plants and its soil is peaty and immature – it is incapable of supporting crops or intensive grazing. The island's thin, patchy vegetation is known as "fellfield" rather than tundra. Both the Falkland Islands and South Georgia support fringes of tussock grass at sea-level.

Antarctic habitats

Every winter, pack ice surrounds the periantarctic islands (within the 50 degree south latitude), bringing their temperatures down to well below freezing point. Such islands have no fringes of tussock grass. At sea level, their richest vegetation

is a sparse fellfield, much poorer in plant life than that found on South Georgia. There are no ferns and only one or two species of flowering plants. During summer, moss, lichens and algae typify the islands' vegetation, forming scant patches of plant growth on the damp ground (not unlike the polar desert communities of the north). Close to the sea, snow fields that persist into late summer often develop striking patches of green and red – these are caused by algae that proliferate in a thin layer on the snow surface.

Island birds

Most periantarctic islands support very few land birds: South Georgia has indigenous pintails and pipits, while Kerguelen is home to ducks and rails; the islands south of New Zealand have acquired parrots and passerines. However, vagrant waders, swans, ducks and other water birds from South America occasionally enter the region – although none survive for longer than a few days. Continental Antarctica has no semblance of tundra vegetation: only bout two per cent of its land surface is ice-free; its soils are poor and undeveloped, and the whole continent supports only

LEFT By far the coldest of the continents, Antarctica has no permanent human inhabitants. In winter, a small number of scientists and supporting staff work at scattered bases; many more visit on summer expeditions, and now there are increasing numbers of tourists. The continent is managed under an international Antarctic Treaty that provides rules and guidelines for the protection of its natural beauty and wildlife.

thin, widely-scattered polar desert vegetation. The richest vegetation grows around the coast, while individual clumps of moss and lichens manage to survive wherever rocks appear in the icebound interior – the smallest trickle of meltwater is sufficient for their growth. Exploration parties have recorded the presence of lichens as far south as bare rocks appear, at altitudes of 7–10,000 ft and even within five degrees (350 mi) of the South Pole.

Since the periantarctic islands were never connected to a continent (at least during the 60 million years of mammalian evolution), they have no mammals of their own. The Falkland Islands proved the only exception with its Falkland Island wild dog – a small, coyote-like animal known to human settlers in the 19th century. Unfortunately, the settlers exterminated the animal, since it preyed on their sheep. Geologically, the Falklands were never part of South America – it is more likely that they once formed part of the African continent. Consequently, they lack the llamas, armadillos and rheas that might otherwise have enlivened their land fauna. However, the islands have acquired more than 40 species of land birds, most of them from Patagonia.

Imported wildlife

Sealers and whalers who hunted the southern oceans in the 19th and early 20th centuries, brought rats and mice with their stores, and landed rabbits, sheep, goats, cattle, pigs and other mammals to provide food for shipwrecked crews. Some of the introduced mammals flourished, but usually to the detriment of the native vegetation. Feral cattle still graze among sea lions on the Auckland Islands; rabbits and wild sheep have devastated parts of Kerguelen; and brown rats forage among the tussock grasses of South Georgia.

In the 1920's, whaling ships transported small stocks of reindeer from Norway to South Georgia to provide hunting game. The reindeer found good feeding grounds on the island's uplands, and soon spread beyond their original settlement areas. The hunting activities of whalers controlled the reindeer population growth for a time, but after whaling ceased in the mid 1960's, the animals were left to themselves. Today, several thousand reindeer inhabit South Georgia – their herds range over several peninsulas on the northern shores of the island.

RIGHT Wandering albatrosses court on a windy hillside in the Bay of Isles, South Georgia. Largest of all the albatrosses, with a wingspan of more than nine feet these birds spend most of their lives in the air. Patrolling the rough seas of the roaring forties and fifties, they feed on fishes and squids. Each pair lays a single egg and incubates it for eight weeks; the chicks take more than a year to mature and reach independence.

The changing Arctic

For thousands of years, the Arctic developed without human intervention. The first humans arrived about 30,000 years ago from central Asia, probably through forest which extended much farther north than it does today. Some 10,000 years ago, they spread across the Bering Strait (then dry land) to western North America, and farther eastward to Greenland. Wearing animal skins and living as nomads in small, scattered, self-sufficient communities, they developed similar but separate cultures. From these travelers arose the Arctic and sub-Arctic peoples we now identify as northern Indians, Inuit (Eskimos), Aleuts, Chukchi, Eveni, Yakuts, Samoyeds and Saami (Lapps).

Always limited in numbers, they adapted to their harsh environment,

having neither the means nor the desire to alter it. The arrival in the 17th and 18th centuries of explorers, sealers and whalers from the south, soon to be followed by trappers, traders, missionaries and administrators, brought irreversible changes to the Arctic way of life. Disease and alcoholism were introduced, together with a destructive dependence on trade goods that entirely altered their relationship with the environment. Guns, knives, patent traps, southern foods, clothing and medicines eased their hard lives, but required payment in furs. While the Arctic people had previously subsisted on only a few natural resources, they later learned to exploit their environment on a large scale to meet the demands of the south.

The problem remains today. Northern people hope to share the benefits of a modern world, but cannot pay their way from the natural resources that earlier supported their simpler way of life.

Few would want to turn back the clock; even in the early 20th century, many Arctic folk died in infancy and survivors seldom lived for more than three decades. Today, industry is learning to mine the Arctic's oil and mineral deposits, and develop tourism in the

ABOVE King penguins, the brightest and most colorful of all penguins, breed in colonies several thousand strong on the beaches of South Georgia and other sub-Antarctic islands. Penguins are found only in the Southern hemisphere, and mainly in the cooler waters. Five species nest on Antarctica and the neighboring islands, and feed in the icy waters.

region. Arctic communities that have political power try to ensure that profits from these ventures ease their own transition, and improve things for their children.

Irreparable damage

The exploitation of the tundra inflicts severe, and sometimes irreparable, damage. Oil rigs and pipelines, spoil-heaps, roads for trucks and coaches, new hotels and airstrips, all scar the landscape.

Unfortunately, the huge areas that have already been altered by development can never be expected to return to their original state. If we value the large national parks and reserves that all Arctic nations have set aside, we must guard them as irreplaceable and beyond price – to be protected from thoughtless exploitation and development.

Antarctica

Antarctica has never had an indigenous human population and has a history of human intervention that dates back less than two centuries. The first expedition huts were built in the final decade of the 19th century. Today, there are some two dozen active research stations, and many more abandoned ones scattered across the continent. Some – those of the early explorers Scott and Shackleton, for example – are treated reverently as historic relics. Those in areas of high precipitation have disappeared without trace under the snow. Many of the more recent stations, established for political or scientific reasons, have quickly decayed into ruins.

Periantarctic islands (those on the Antarctic boundary) are owned by France, Britain, Australia, New Zealand and other nations, who care for them with varying degrees of responsibility. Antarctica and its neighboring islands up to 60 degrees south are protected under the international Antarctic Treaty. All nations active in Antarctica have signed the Treaty, and have agreed to the additional conservation measures prescribed by it – the measures are effective both within and beyond the treaty area.

Conservationist concerns

Many ecologists, conservationists and others with polar interests have serious doubts about the Treaty's ability to deal with growing conservation problems in the area. They are especially concerned with hazards that are likely to arise from mining and drilling for oil; some wish to see Antarctica excluded forever from all threats of mineral extraction.

Even in the depths of winter when deep snow cloaks the peaks of North America, the Rocky Mountain goat haunts the bleak mountaintops and precipitous crags. It has a double-layered coat of thick fur that keeps the winter chill at bay.

MOUNTAINS

Shaped by colossal forces, mountains are the most awesome of natural features. As such, they offer a dramatic range of habitats – from the forest-cloaked foothills to the bare, rugged heights where chamois leap over chasms, snow leopards ambush prey and golden eagles soar overhead

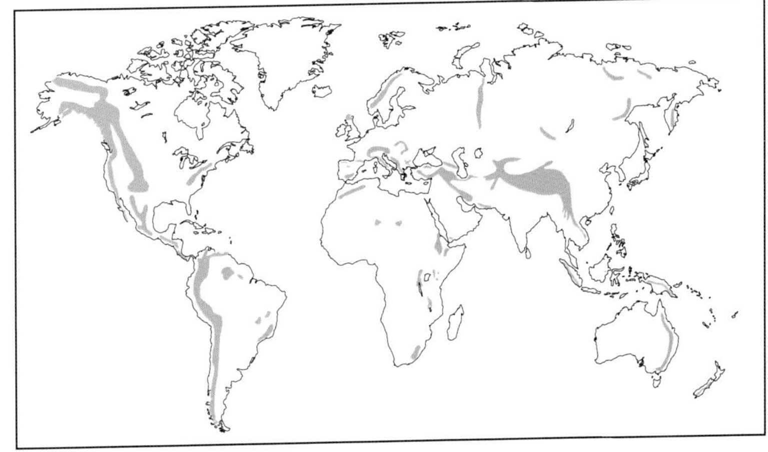

ABOVE The raven inhabits wild, rugged habitats and is a common sight on the high mountain slopes, where its acrobatic courtship flights provide a dramatic spectacle among the crags. In the Himalayas, it nests at heights of up to 17,000 ft above sea level. Both a predator and a scavenger, it scours the slopes for rodents and young birds, and will tear flesh from the carcasses of dead animals, including domestic livestock. PAGES 302–303 Life at high altitude places great demands on mountain wildlife. Thin air, fierce winds, cold nights and strong sunshine combine with treacherous slopes and poor soils. Only hardy animals, such as the ibex, range up towards the highest summits where conditions are harshest. Small bands of ibex survive in the heights of Eurasia, browsing on the scant vegetation, even when snow covers the ground.

At 20,700 ft, Chimborazo is one of the highest mountains in the tropics. To the west of the peak, the land falls away to luxuriant lowland forests and banana plantations, while to the east, across a belt of lesser mountains, lies the vast tropical rain forest of Amazonia. Chimborazo's high summit gleams white with a permanent cap of ice and snow and glaciers extend down-slope for a thousand metres or more.

The chill of altitude

The factors that make conditions on Chimborazo so different from those in the nearby equatorial lowlands recur throughout the world wherever mountains rise far above the landscape. Rapid increases in altitude bring marked differences in climate. Temperatures drop about one degree for every 500 ft of ascent, so that a relatively low mountain that rises 5,000 ft from its foot will be 10 C (18 F) cooler than the surrounding plain. Night-time frosts occur at altitudes of 5,000 ft even in tropical regions, especially if cloudless nights allow heat to escape easily from the atmosphere. At higher altitudes, temperatures plummet to well below freezing point. In the Himalayas – the mountain range that contains the greatest peaks in the world – temperatures at the summits are often as low as −40 F.

Coolness increases with altitude because the density of air steadily declines with increasing height. The thinner the air, the less heat energy it can absorb from the sun's rays. Air pressure at 20,000 ft may be half that at sea-level.

Exposure to the sun

Low air pressure has other harsh effects on the mountain environment. The midday sun brings intense sunshine to the landscape, especially in the tropics. Since the thin air traps little of the incoming radiation, the ground surface and the plants and animals that dwell on the mountain receive the full force of the sun. Although the surrounding air may be cool, strong sunshine causes problems of overheating and exposure to ultraviolet radiation. When night falls, the thin air traps little of the heat radiating back into space, and the ground temperature falls to several degrees below freezing. Animals experience breathing difficulties in the thin mountain air, because the

ABOVE The Himalayas are the greatest of all the mountain ranges, with a vast chain of saw-toothed peaks towering to heights of more than 27,000 ft. The icebound summits are among the most inhospitable places on earth, with hurricane-force winds sweeping their slopes and temperatures falling as low as −40 F.

supply of oxygen is low. Above 10,000 ft, the shortage of oxygen can cause severe sickness in humans and animals not used to life at high altitudes.

The highest slopes

Conditions on the highest slopes of the mountains are the most testing for wildlife. There the air is at its thinnest, and the exposed heights experience some of the fiercest gales on earth. Hurricane-force winds regularly sweep across the snow fields of the Himalayas. The strongest surface wind ever recorded blew at a speed of 200 mph across Mount Washington in the USA. Steep, rocky slopes near the mountains' summits make it difficult for animals to move about, and soils on the slopes are thin because particles are easily washed or blown down to lower levels.

Only a few specialized animals can survive on the high mountain-tops, but lower down, where conditions are more favorable, highland areas support abundant wildlife. The varying nature of the terrain – from flat basins and rolling slopes to steep-sided valleys and vertical crags – combines with the steady change in climate up the mountain to provide a variety of habitats for wildlife. A journey to the top of a mountain therefore reveals a much greater variety of plants and animals than a trip across a lowland plain.

A variety of mountain types

Mountains exist throughout the world wherever the land rises abruptly above the surrounding terrain. However, countries differ greatly in their definition of the term "mountain." In the USA, most mountains are about 10,000 ft, but in Nepal, a summit twice that height is merely a "foothill" of the mighty Himalayas. Mountains also differ ggreatly in form. The Scottish Highlands are low and rounded when compared with the steep, jagged skylines of the Himalayas and the Alps. Mountains usually occur in long chains or ranges, but isolated massifs sometimes appear. The great volcanoes of East Africa rise straight from the surrounding plains, separated from one another by hundreds of miles of savannah. Some mountains in arid regions are bare masses of rock, while the slopes of others are covered with dense forests.

Treeline and snowline

The treeline – the altitude at which trees cease to grow – varies from one mountain range to another according to the climate. In Scandinavia, close to the Arctic Circle, coniferous forests reach altitudes of only 3,000 ft before the average daily temperature drops too low for growth. In the tropical warmth of Java, however, trees grow on the slopes of volcanoes at altitudes of as much as 9,000 ft. The snowline – the level at which snow remains frozen throughout the year – shows a similar pattern. In the Scottish highlands, permanent snow fields linger on shaded slopes

at 4,000 ft, but, in the tropics, they occur only at the summits of 20,000 ft high peaks. In winter, deep snow extends well below the snow-line in the mountain ranges of the Alps, the Himalayas and the Rockies.

Ranges around the world

Every continent has its major mountain ranges. Northern Europe's ranges include the Scandinavian Uplands that stretch the length of Norway, and the ancient, worn slopes of the Scottish Highlands. To the east, the long north-south chain of the Urals divides European USSR from Siberia, while a chain of ranges begins in the south, stretching across the breadth of Eurasia. In southern and central Europe, such mountain ranges include the Pyrenees, the Alps, the Apennines and the Carpathians. The Alps, which lie in France, Switzerland, Austria and Italy, have several peaks of over 11,000 ft in height. They reach their highest point at Mont Blanc – 15,000 ft above sea-level. Eastwards, the mountains continue with the Taurus range of Turkey and the barren Zagros mountains of Iran. The Caucasus, to the north, stretch from the Black Sea to the Caspian Sea in the USSR. They exceed the Alps in height; their tallest peak is El'brus (18,500 ft).

LEFT Conifers occur higher up the mountainsides than broad-leaved trees since they are better able to withstand the cold. Belts of coniferous forest girdle the peaks of many mountain ranges, particularly in temperate and northern regions. Here, a larch wood grows on a middle slope in the Alps. Unlike most conifers, larch lose their needles for the winter, bringing fall colors to the Alpine forests.

Asia contains the highest of all the world's ranges, the Himalayas. Rising from the plain of the Ganges, the mountains form a vast ice-capped wall across the northern fringes of the Indian subcontinent. To the north, the highlands continue as the vast plateau of Tibet, while to the west lie massive offshoots of the Himalayas – the Karakorams, the Pamirs, the Hindu Kush and the Tien Shan mountains – all with peaks of well over 23,000 ft in height. The Himalayan region of India, Nepal, Sikkim, Bhutan, Pakistan, USSR and China contains the world's highest peaks. The greatest of all, Everest, soars to 29,000 ft on the border of Nepal and China.

North of the Himalayas lie the high Altai mountains of Mongolia and the USSR and a series of lesser ranges in eastern Siberia. To the south, the mountain chain runs into South-east Asia and Indonesia, with many volcanoes on Java and the surrounding islands. New Guinea contains a string of mountains over 12,000 ft in height and New Zealand's Southern Alps peak at over 10,000 ft; Australia contains few mountains for its size. Though the Great Dividing Range runs the length of the eastern seaboard, most of its peaks are below 5,000 ft in height.

African and American peaks

Africa, like Australia, is not a mountainous continent. Its highest peaks, Kilimanjaro (19,000 ft) and Mount Kenya (17,000 ft), are both isolated East African volcanoes, and its major mountain ranges cover only a small proportion of the land. North Africa's better-known ranges are the Atlas mountains of Morocco

and Algeria, and the Hoggar and Tibesti massifs of the Sahara. The Ethiopian High-lands are the most extensive range, covering most of Ethiopia; narrow chains of the same range run south along the Rift Valley and comprise the Aberdare, Ruwenzori and Virunga ranges in East Africa. In southern Africa, a chain of steep-sided peaks make up the Drakensberg range.

A spine of high mountains runs the length of North and South America from Alaska to Tierra del Fuego. Mount Mckinley (20,230 ft) in the Alaska range is the highest mountain in North America; a line of peaks continues south through the Coast Ranges of Canada to the Cascade and Sierra Nevada ranges of the USA. The Rockies of Canada and the USA form a broad parallel range to the east. The Sierras of Mexico continue the mountain spine into Central America, and in South America it reappears as the Andes. The Andes, which hug the coastline from Colombia to Chile, come second only to the Himalayas in height. Many of their peaks top 20,000 ft, and the highest, Aconcagua, reaches 22,800 ft. Other, lower ranges in the New World include the appalachians of eastern USA, the Guiana Highlands, which run from Venezuela towards the mouth of the Amazon, and the Brazilian Highlands of south-eastern Brazil.

ABOVE Different rock types and rainfall patterns create a variety of mountain landscapes around the world. The barren, semi-desert landscape of Amba Alage in Ethiopia contrasts with the lush forest that coats the lower slopes of mountains in humid regions of the tropics.

Several oceanic islands have steep mountains. Often, they are the summits of giant submarine mountains that rise from the sea-bed. The volcanoes Mauna Loa and Mauna Kea that dominate Hawaii reach over 12,000 ft above sea-level, but their bases are so far underwater that they are actually equivalent to Mount Everest in their true height.

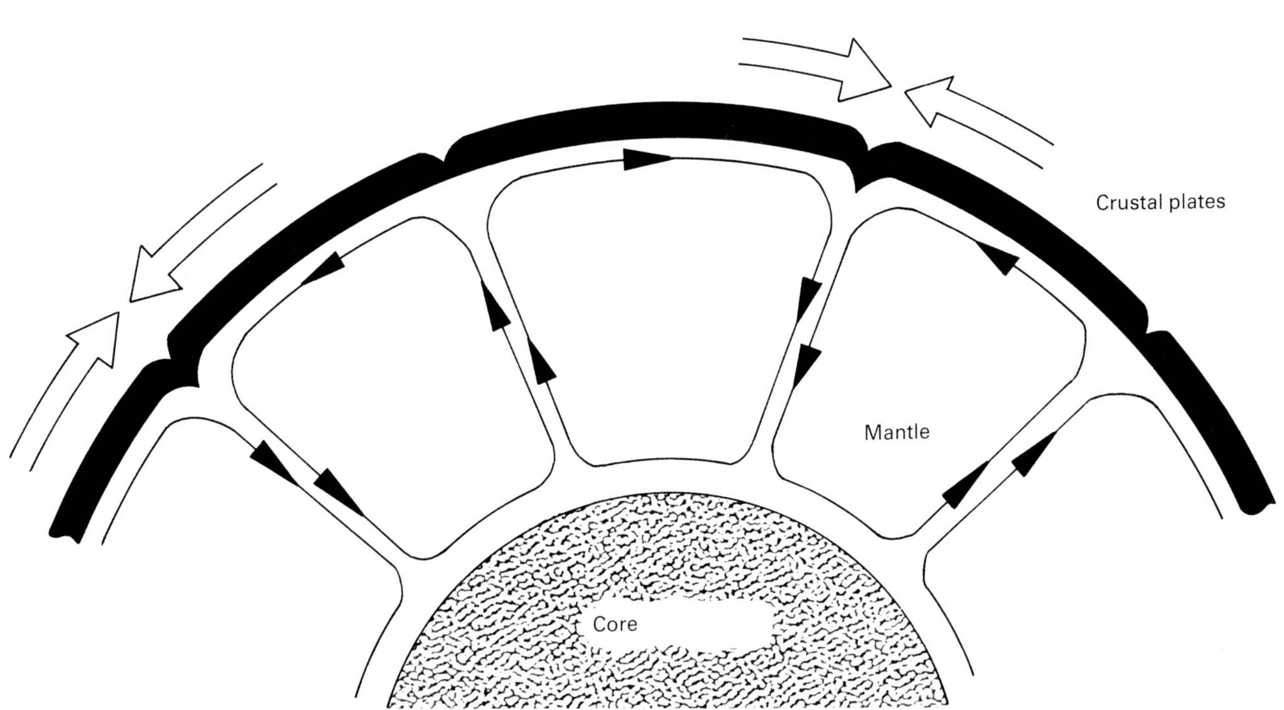

ABOVE Mountains result from great forces at work beneath the surface of the earth. The interior of the planet comprises an immensely hot core, surrounded by a thick mantle and a thin outer crust. The slow circulation of hot, compressed rock in the mantle pushes great sections of the crust, called plates, across the surface. When adjacent plates collide, immense pressure causes the rock to buckle, throwing up mountain ranges. The strain on the rock also produces weak points in the crust, from which volcanoes emerge.

Rise and fall

Though mountains appear to be vast, static objects, they undergo a continual process of change. Some, like the Himalayas, are steadily increasing in height. The rate of change is imperceptible without the aid of sophisticated instruments, since the mountains grow a fraction of an inch every year. Over the course of millions of years, however, they gain thousands of feet and new peaks can rise from land that was once merely a lowland plain. Other mountains, like the Scottish Highlands, that have long stopped growing, are now gradually shrinking. Piece by piece, fragments of rock break free from their summits and fall down-hill. Millions of year after their birth, mountains finally disappear, receding back into the lowlands.

The titanic forces that shape mountains and thrust them skyward originate in the interior of the earth. Beneath the continental land masses and the ocean bed that together form the earth's crust, lie thousands of miles of solid rock known as the mantle. The mantle encloses the core – an immense ball at the earth's center composed partly of molten metal. Hundreds of miles below the surface of the earth, the mantle is intensely hot and under immense pressure from the weight of rock above it. Geologists believe that it exists in a plastic, or semi-liquid, state, in which currents of solid rock can slide past one another, albeit at a minute rate.

Differences in temperature between cooler rocks near the surface and hot, inner rock lead to a pattern of convection currents within the mantle. Over millions of years, "plastic rock" circulates upwards to the surface, sideways beneath the crust and back down into the interior (see diagram above). As the currents flow near the surface, they move great sections of crust called plates. Where convection currents converge, and flow back downwards, the plates meet and thrust up mountains.

Building mountains

Collisions between plates set up tensions in the crust that warp and fracture the rock. Warping produces folds in the rock layers

LEFT Permanent fields of ice and snow cap the summit of Pico Bolivar. Measuring 15,000 ft high, it is the highest peak in Venezuela. Steep, jagged rock faces are typical of snowbound summits, where ice is a powerful agent of erosion. Glaciers gouge out deep hollows in the rock, and when rainwater collects in fissures, it freezes and expands, shattering stone into sharp fragments.

that push peaks skywards. Fractures cause sections of rock to slide against one another along fault lines. If one rock section moves upward, it forms a steep mountain slope that separates the two blocks. Some of the highest mountain ranges occur alongside collision zones, where they mark the boundaries of adjacent plates. India rests on a separate plate from most of Eurasia, and edges northward at a rate of one inch per year. The Himalayas mark the line where the two land masses meet – they are steadily folding upwards with the force of the collision.

Volcanoes form where molten rock from the earth's interior spills onto the surface. Many arise at the boundaries of plates where convection currents drag slabs of crust down-wards, generating great quantities of heat. Some rock material melts and forces its way upward through weak points in the crust. Periodic eruptions bring streams of lava to the surface. After each eruption, the lava cools and solidifies, eventually building up a cone of rock with a central crater. There are some 10,000 volcanoes around the world, several hundred of which are active. The volcanoes in North and South America, Asia and Oceania form a chain of cones that almost encircles the Pacific Ocean.

Wear and decay

Once mountains appear, the elements immediately combine to wear them down. Streams gather rainfall and erode furrows in the rock as they descend in torrents down slopes. Where they come together, deep, V-shaped valleys sweep down the lower slopes. Water that seeps into the soil collects chemicals that also act on the underlying rock, gradually breaking it down through a process of decay. On the steep slopes, the soil and rock particles steadily slip and creep downwards, washing into streams that carry the material far away from its source.

Rainfall and rivers gradually reduce the height of mountains, but the action of ice produces more dramatic results. On the cold summits, water that collects in fissures freezes and expands, breaking jagged pieces of rock from the slopes. Where ice collects to form glaciers (slow-moving rivers of ice), immense eroding forces come into play: they dig circular hollows (corries or cirques) at their source, carve mountain peaks into sharp pinnacles and gouge deep, flat-bottomed valleys into the mountainsides.

Europe under ice

Many European mountains show evidence of glaciation, though no

glaciers are present in the landscape today. Britains's broad valleys and circular corries developed long ago, during a series of Ice Ages that dramatically influenced the landscape of the Northern hemisphere.

With the coming of each Ice Age, the polar ice sheet advanced southwards, bringing glaciers to the Scottish Highlands. A lowering in temperature brought freezing winter conditions to much of Europe, causing tundra and conifer forests to replace the broad-leaved woodlands of the lowlands. The southward spread of habitats from northern regions also led to changes in the distribution of wildlife. Species accustomed to temperate climates disappeared from northern Europe, while cold climate animals took their place as they increased their range southwards.

Seeking cold conditions

When the ice sheets receded at the end of the last Ice Age, temperate species returned and northern species retreated. The retreat was not complete, however: mountains, with colder weather than the surrounding lowlands, could still provide a refuge for animals adapted to the chill. As warm conditions returned, many species climbed higher into the mountains. Eventually, they became isolated from their northern counterparts and formed separate populations among the peaks. The ptarmigans are a relic population in many mountainous areas of Europe, including the Alps, the Pyreness and the Scottish High-lands. The largest populations, however, occur in the tundra zone of Arctic Eurasia and North America.

ABOVE Mottled coloring provides the patarmigan with excellent camouflage on the high slopes in summer and the fall. But when winter sets in and a blanket of snow covers the mountaintops, the ptarmigan molts. It replaces its mottled plumage with pure white feathers to match its surroundings. Ptarmigan spread south from northern Eurasia during the Ice Ages, when much of Europe was beset by the cold conditions. When the ice sheets began to retreat some 12,000 years ago, the birds moved with them. But they left behind relic populations high in the Scottish Highlands, the Alps and the Pyrenees where cold conditions still prevail.

Rare mountain animals

Marooned relics from the Ice Ages do not account for all the wildlife of the mountains. Many creatures are adapted exclusively to life on the high slopes. Some are upland races of species that

WINGS IN THE WIND

Across the mountains of Eurasia and North America, the golden eagle reigns supreme as the ultimate aerial predator. Though it occurs in a variety of wild habitats, the mountains are its true stronghold, and it is more at home in the bleak, windswept terrain above the treeline than any other eagle.

The majestic golden eagle is a huge brown bird – the wingspan of a female may measure more than 7 ft. On outstretched wings, it can soar for hours in the violent winds that buffet the peaks. With wings partly bent, it can glide through the air at speeds of 125 mph. In the breeding season, golden eagles perform breathtaking dives and upward swoops as they court their partners.

Methods of attack

The golden eagle sometimes dives onto prey, striking large flying birds, such as geese or ducks, in the manner of the peregrine falcon. However, it usually hunts close to the ground, hugging the contours of the land. Often it will fly just behind a ridge, controlling its movements with delicate adjustments of its wingtips, rising from time to time to survey the land. If it sees a marmot or a grouse, it will sweep over the ridge and swoop down with its talons forward, taking its victim by surprise. The eagle has the agility to follow the darting, zig-zagging run of a hare, and it will kill its prey outright with the force of a single blow. Small mammals and game birds are the golden eagle's main prey, but young ibex and deer calves are also vulnerable.

BELOW One of the most powerful of all mountain predators, the golden eagle hunts across vast stretches of mountain where it occupies a home range of about 45 square miles. It surprises its prey in the open and races forward with its talons extended. At the moment of impact, the bird may be traveling at 80 miles per hour.

ABOVE Two male ibex battle for dominance in the Italian Alps, interlocking their thick, curved horns in a show of strength. Ibex spend most of their lives high above the treeline, foraging and breeding among the alpine meadows and crags. They seldom descend to the lower slopes and valleys, even in winter. In summer, they climb close to the permanent snowfields, where temperatures remain cool and there is little competition for frest plant growth.

normally inhabit the lowlands. For example, guerezas are colobus monkeys of the forest canopy that dwell in many parts of East Africa. Those that inhabit the cool mountain forests and bamboo groves on the slopes of the Aberdares have a more luxurious coat of fur than their lowland cousins.

Other animals are true mountain species, closely adapted to the upland environment and more at home among the peaks than in any other habitat. Several species of goat antelopes live only at high altitudes, where their thick coats and sure-footedness equip them well for life on the crags. The chamois, the gorals, the ibex and the Rocky Mountain goat all keep to the highland zone. The ibex occurs in high spots throughout Eurasia, from the Alps through the Middle East to the Himalayas and the Altai mountains; a small population also survives in the mountains of Ethiopia. Many birds of prey are closely associated with mountain habitats – they include the Andean condor, the lammergeier and the golden eagle (see box on page 312). Snowcocks are almost exclusively upland birds.

Land-locked islands

The separation of mountain ranges from one another, each surrounded by stretches of lowland, creates an ecological situation comparable to that of islands in the oceans. In the same way that island animals are isolated by the barrier of the sea, mountain creatures may become isolated by their adaptation to high altitude. Many species cannot survive in the lowlands long enough to cross to other highland regions. The vole *Dolomys bogdanovi*, for example, lives only in the craggy limestone peaks of Yugoslavia, where it forages for plant matter among the rocks, storing excess food in its nest for the winter months.

Isolation can lead to the rapid evolution of different species and subspecies of certain families. The forested slopes of the Appalachians are home to no fewer than 27 species of woodland salamanders, many of which have distinct sub-species confined to different ranges. The Cheat Mountain salamander and the Roanoke salamander both live in small areas within the Allegheny range.

Bands of life

One common feature of the mountain environment throughout the world is the arrangement of different habitats in layers according to altitude. The declining temperature creates bands of vegetation, each containing plants more resistant to cold than those in the previous band. Often, sharp boundaries separate each layer, creating a distinct pattern visible from many miles distant. Dense forests typically cloak the lower slopes, but grade into scrub and meadows, and finally into a tundra-like zone beneath the ice caps of the highest peaks. The layered pattern is at its most extreme in high tropical mountains. Hot, lush forests at their feet contrast sharply with the freezing habitats of the summits.

Similar climatic conditions prompt the growth of similar types of plants across many separate mountain regions. A conifer forest at the

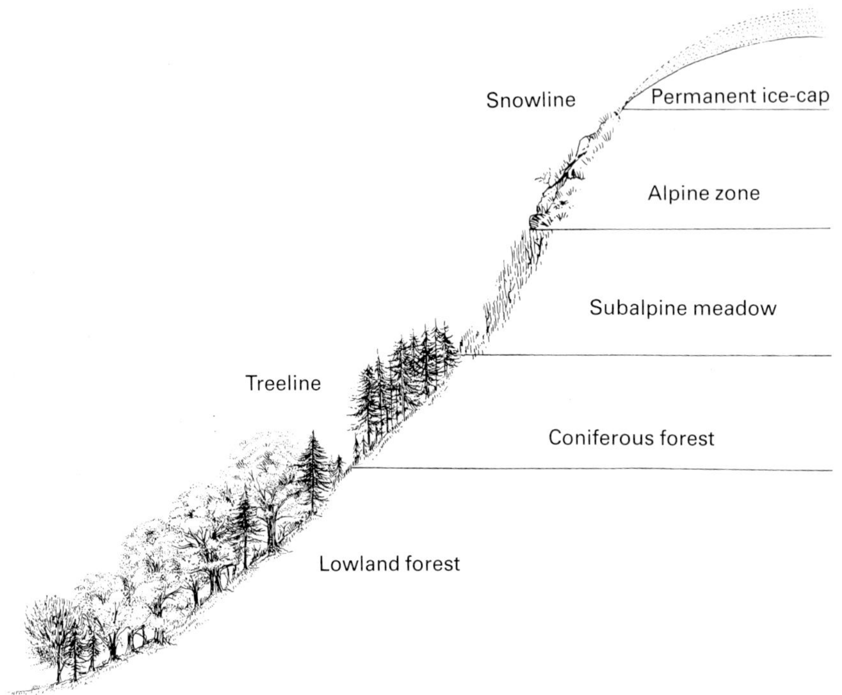

base of a Scandinavian mountain has its equivalent as a layer on an alpine mountain five hundred miles to the south. Since the highest of the Alps have glaciers at their summits, a climb from the valley bottom to the top is similar to a journey through the vegetation zones from central Europe to the North Pole. The different habitats across 3,000 miles of lowland are condensed into a vertical distance of 15,000 feet up a mountain.

Stealing the rain

Rainfall combines with cold to influence the types of plants that grow in each vegetation band. The upwind slopes of mountains often receive heavy rainfall because the rising ground forces air high into the sky where it cools to form rain clouds. The wettest place on earth is the mountain Wai-'ale-'ale in Hawaii, which receives an average yearly rainfall of over 36 feet. By the time the air currents reach the slopes on the far side, particularly of broad ranges, they may have shed most of their moisture. As they descend, they become warm and dry, bringing contrasing aridity

ABOVE The changing climate up mountain slopes creates a distinct pattern of vegetation from the base to the summit. In temperate regions, lowland forest merges with coniferous forest. Higher up, the conifers cease, and subalpine meadows and alpine tundra appear. A permanent snowline exists on the highest mountains. Above the snowline, ice and snow dominate the landscape throughout the year.
RIGHT Luxuriant mountain forest spreads up steep mountain slopes on the Caribbean island of Guadeloupe. Similar to the lowland rain forest, but with a lower tree canopy, the mountain forest supports a rich wildlife community of birds, frogs and butterflies.

to the downwind slopes. The phenomenon – known as the rain-shadow effect – sometimes forms deserts in the lee of mountains and may cause the habitat bands on the downwind slopes to be quite different from those upwind. In the Elburz mountains of northern Iran, the forested northerly slopes receive over 6 feet of rain per year, while the thinly-vegetated southern slopes experience desert conditions.

Exploiting different habitats

The layers of vegetation influence the distribution of wildlife, providing different habitats in close proximity and a wide variety of food sources up the mountain slopes. Some mountain animals wander from one band to another in order to exploit the different habitats. Deer travel from the cover of forest to forage on the open moors above, and buzzards sweep over woodland, meadows and crags in search of prey. Other animals are more closely tied to a particular habitat band, and seldom stray above or below. Most of the salamanders of the Appalachians, for example, keep to moist woodland on the middle slopes. In the Caucasus, the Caucasian snowcock lives on rocky terrain and meadows above 8,000 ft through all but the harshest winter weather. Another game bird, the Georgian black grouse, inhabits the forested belt below, rarely straying into the snowcock's domain.

Plant life upslope

On mountains with well-defined vegetation layers, the pattern of plant life shows broad similarties throughout the world, with a general reduction in the density of vegetation up the slopes, particularly above the treeline.

Forest at the foot

In humid parts of the world, forests usually form the natural vegetation at the base of mountains (although in many regions, cultivation of the lowlands has stripped away the original habitat). In tropical and subtropical regions, lowland rain forests and monsoon forests grade into evergreen mountain forest. Mountain forests are characterized by humid conditions and a lower tree canopy on areas of rising ground. Where cloud immerses the slopes or where cool air collects and shrouds the trees in mist and fog, cloud forest develops. Here, mosses drape the stunted trees, and murky conditions hamper visibility.

Rhododendron groves

In Africa, the pure mountain forest belt often ends abruptly with the appearance of dense stands of bamboo that run to the limit of forest growth at about 10,000 ft. In the Himalayas, the mountain forest grades into rhododendron groves that grow to 30 ft in height; the rhododendrons produce a distincitve type of forest, with tough, glossy leaves. Higher up, the rhododendron groves merge with conifers that comprise the forests that grow up to the treeline.

Further north, where temperatures start at a lower level, temperate broad-leaved woodland and mixed woodland clothe the lower slopes of mountains. After several hundred feet, the woodlands give way to pure coniferous forests. Still further north, conifers, such as spruce, continue up the lower slopes from

the great boreal forests on the lowlands. Here, the treeline is usually no higher than 3,000 ft.

Stunted growth

Some mountain conifers reach extraordinary size – at nearly 300 ft in height and circumference around the base, the giant sequoias on the lower slopes of the Sierra Nevada are among the greatest of all trees. Towards the treeline, however, conifers typically become strangely stunted and twisted. The effects of cold, wind and shallow soil combine to reduce the height of trees to no more than a few feet. On exposed, windswept parts of northern mountains, trees sometimes form small, ground-hugging thickets called 'krummholz'. In the tropics, sections of mountain forest in windswept locations decline in height to little more than that of a human, becoming deformed elfin woodland.

In drire regions of the world, the lower slopes of mountains may bear no trees at all. In the foothills of the western Pamirs, the cold desert landscape of lowland central Asia blends with an open habitat of scattered bushes and grasses. The same habitat continues to the altitude at which the treeline would run on more humid mountains.

Towards the top

The landscape above the treeline is open with sparse foliage. Two zones of vegetation – the subalpine and the alpine zones – often form distinctive bands across mountain slopes. In humid northern mountains, such as the Caucasus and the Altai ranges, subalpine meadows appear; they have thick carpets of grass and colorful flowers, such as

LEFT The hardy Swiss stone pine grows high in the Alps, sometimes appearing at heights of 10,000 ft. But though it is specialized for altitude, it cannot reach the greatest peaks. Increasing cold and thinning soils limit the height at which trees can survive, and on many mountains there is a well-defined boundary called the treeline. The altitude of the treeline varies greatly around the world. In northern latitudes, it may be lower than 3,000 ft while in the warmer tropics, dense forests may grow at 10,000 ft.
RIGHT Above the treeline in the Alps, low-growing plants sprout in boulder-strewn meadows. Grasses and herbs predominate, providing food for grazing animals such as deer and goat antelopes. Alpine marmots spend their lives in the meadows, gnawing the shoots and digging extensive burrows.

giant buttercups and umbellifers. In the Tient Shan, honeysuckles and dwarf junipers extend above the conifer forests, while in the dry Pamirs, the subalpine zone resembles the steppe grassland habitat of the USSR.

In Africa, the subalpine band of vegetation takes on a strange appearance. Misty, high-altitude moorland lies above the bamboo groves on Mount Kenya. The area supports sedges, tussock grasses and huge, tree-like heath plants – giant relatives of the plants that grow on many north European heaths. Dense growths of mosses and lichens cover the latter's twisted stems and branches.

In the alpine zone, the plant life is subject to the extreme effects of high altitude. Towards the permanent snowline, the landscape is similar to the tundra of the Arctic. Though permafrost is rare and the ground is steep with rocky outcrops, ground-hugging vegetation, bogs and icy pools dominate the ground surface. Alpine meadows often produce brightly colored blooms in spring – alpine poppies, violets and gentians compete to attract the few insects that can survive the coldness of high altitude.

Cushion-shaped plants

The intense chill of night, combined with the scorching mid-day sunshine, the high winds and the thin soil, has prompted some ingenious adaptations among plants of the alpine zone.

To reduce their surface area against heat loss and moisture loss, and to keep delicate parts out of the wind, many alpine plants grow flat against the ground. They often form a rounded cushion-shape, as in the prickly thrift and the rock jasmine. Though they have short or virtually absent stems, the plants may have deep tap roots to hold them fast in the wind, and a mass of smaller roots to extract nutrients and water from the scant soil. The tiny leaves pack tightly together – often as bunches of rosettes – and overlap to form a hard, compact mound. In the intense cold, chemical reactions happen slowly and the

processes that build new plant tissue are labored. a cushion plant may add only one or two leaves each growing season, and may take decades before it is ready to flower. Nevertheless, given time, they can develop into structures known as "vegetable sheep" in the Andes and in New Zealand's Southern Alps.

Insulation from the cold is a common feature of alpine plants. The leaves of the edelweiss of the Alps have a furry coat of white hairs that traps warm air when the night cools the atmosphere. In the saussureas of the Himalayas, the plants are entirely covered with a thick fuzz of hairs that obscures the pattern of stems and leaves. An opening at the top allows insects inside to pollinate the plant, and the insulation provided by the furry coat keeps the plant warm enough to provide a night-time shelter for bumble bees.

Pale hairs and white deposits of lime on the surfaces of leaves reflect excessive midday sunshine from the plants, including the harmful ultraviolet rays that can severely damage plant cells. Plants such as the common alpenrose limit moisture loss by having scales around the pores of their waxy leaves. Alpine plants continually struggle to draw moisture from the thin mountain soil, especially during the freezing winters.

Protective tussocks

Alpine grasses usually develop in tussocks (tufts or clumps of grass) on the alpine meadows. They grow close together in order to shield one another with their leaves. When the leaves and leaf-bases die, they form a slowly-decaying, felt-like mass

TOP LEFT When flowers, such as gentians, come into bloom in the spring, the alpine meadows of the north burst with color. Short stems and bunched growth reduce the surface area of the plants, preventing heat and moisture loss. It also protects them from the high winds.
LEFT The white, furry hairs on the leaves of edelweiss act as insulation against the chill of high-altitude. The hairs trap a thin layer of air around the leaves, reducing heat loss when the temperature plunges at night.
ABOVE Strange, giant plants grow on the high moors in equatorial African mountains. At 12,000 ft in the Ruwenzori range, tree groundsels produce rounded bunches of leaves up to 20 feet from the ground. Giant tree heaths, festooned with moss, grow several feet higher.

around the bottom of the tussock. The felt soaks up moisture and insulates the new shoots and buds at the base of the grass. Sunbirds on Mount Kenya build their nests in tussocks, where the insulation effect reduces the chance of night-time cold destroying their eggs. Tests on grass tussocks more than 12,000 ft high in the African mountains reveal that the temperature of the interior may remain above freezing while the air temperature falls below freezing.

Some mountain plants generate enough heat through their own chemical processes to warm the ground near them. In spring in the Alps, soldanellas and crocuses remain a few degrees warmer than their surroundings. The warmth is sufficient to melt holes in the receding snows, through which the flowers can open. Night-time frosts in the East African peaks cause needles of ice to form in the soil; as the ice hardens, it pushes the surface of the soil upwards. The daily soil motion would harm delicate plants, but the diminutive *Subularia* plant generates so much warmth that ice cannot form around it. As the bulk of the soil heaves upwards in the night, the plant and the soil close by stay in place.

Tropical giants

A few specis of alpine plants grow to an enormous size, unlike most of the neighboring plantlife which keeps close to the ground surface. In Africa, tree groundsels and giant lobelias tower 18 feet into the air – many times taller than the groundsels and lobelias from other hibitats. In the Andes, a relative of the groundsels grows to some 30 feet in height, and the remarkable rocket-shaped flower-spike of *Puya raimondii* – a relative of the pineapple-thrusts upwards to over 35 feet above the ground. Botanists do not yet understand why the plants have responded to the adverse alpine habitat by turning into giants.

Hairs insulate the leaves of tree groundsels, and some of the giant lobelias grow long, narrow leaf bracts that give the column-shaped plant a woolly appearance. By breaking up the flow of air around the lobelias, the leaf bracts reduce the loss of heat at night. When the leaves of tree groundsels die, they do not fall from the plant, but droop alongside the stem. As more

ABOVE The brown bear of Eurasia and North America is one of many lowland animals that regularly venture up the forested mountain slopes. For the European brown bear, however, the mountains have become a last refuge. Once widespread throughout the lowlands of Europe, it has been driven from most regions by centuries of forest clearance and hunting.

leaves die, they pack tightly together to form several layers of insulating material; often, the dead leaf cover is thicker than the stem itself. The insulating cover prevents the plant sap from freezing and ensures that when the outside temperature is two degrees below freezing, the temperature in the interior remains at about 38 F.

Protecting the shoots

The most vulnerable parts of the giant alpines are the shoots and developing flower heads – African alpine plants take special care to protect them from the chill of night. At sunset, the leaves of tree groundsels close up tightly around the flower-heads to form buds. Some lobelias have rosetted leaves that surround a central shoot. The leaves fit together so tightly that they form a bowl that fills with water until it immerses the shoot. At night, sub-zero temperatures cause the surface of the water to freeze over. Just as a layer of ice on a garden pond prevents the water beneath from freezing in all but the harshest weather, the ice in the lobelia's bowl insulates the water and the shoot below. Through the night, they remain one or two degrees above freezing. In the daytime, the ice melts, but the strong equatorial sunshine does not evaporate the fluid because the water is produced by the plant itself and contains pectin – a substance that reduces the rate of evaporation.

Snow-bound algae

In the highest mountains, there is a further habitat above the alpine layer, where life barely exists at all. Above 15,000 ft, plants become increasingly scarce. In the Alps and the Rockies, 10,000 ft is the normal limit for plant growth. The highest ferns have been recorded at 17,000 ft in the Himalayas, while the highest recorded flowering plant *Christolea himalayensis* grew at an altitude of 19,000 ft. Lichens and mosses cling to rocks at even higher levels, but for the most part, the landscape is one of ice and snow.

Algae are the only plants that can survive on the permanent snows. The tiny unicellular plants stain the snow pink-red in parts of the Alps, the Himalayas and the Andes. The red pigment may be a chemical filter that reduces the impact of ultra-violet light passing through the clear air above the peaks. Nourished by sunlight, the algae develop just below the surface of the snow, which protects them from the gales that sweep over the summits. The plants contain a form of anti-freeze that keeps the cell contents fluid.

Mountain wildlife

Animal life in the mountains follows the trend set by the plant life – the total number of species declines with increasing altitude. Wildlife flourishes on the lower slopes, where conditions are mild and food sources varied. Above the treeline, the subalpine and alpine habitats are more the realm of specialists. Since food is thin on the ground, the ability to forage over a wide area is a great advantage – this is one reason that the mountain tops are rich in birdlife, especially the large birds of

prey that soar over the slopes in search of food.

Many of the animals of the lower mountain slopes are hardy lowland species that cope with life in the foothills without the need for special adaptations. Indian rhinoceroses and tigers inhabit the foothills of the Himalayas, while langurs range into the mountains' rhododendron groves. In the Rockies, wapiti and black-tailed deer wander up to the treeline. African elephants roam the mountain forests of East Africa, where they sometimes walk up steep slopes to feed among the bamboo groves and tree heaths. Brown bears frequent mountain forests over much of North America and the USSR – with the clearance of almost all the lowland forests in Europe, mountains are their last refuge. Small populations of brown bears survive in the Italian Alps, the central Apennines, the Pyrenees and in Spain's Cantabrian mountains.

Animals of the lower slopes

Several animal species occur only in mountainous areas, where they spend much of their time on the lower slopes. Most species of the woodland salamanders are restricted to mountain forests. Temninck's tratgopan is one of the many pheasants and their relatives that inhabit the rhododendron belt of the Himalayan foothills. The red panda frequents the conifers further upslope, seldom straying below 5,000 ft altitude. The Ethiopian Highlands are home to several species absent from other parts of Africa. The mountain nyala, which browses among the remaining freg-ments of forest in the mountains, has a shaggy coat to protect it from the cold mountain nights. Its hair is markedly longer than that of its lowland relative, the nyala. Another resident of the Ethiopian Highlands, the gelada baboon, also has long, thick fur. It is well adapted to life on the dry, rocky slopes typical of most of the mountain range; at night, the gelada shelters in caves or under cliff ledges.

Fighting the chill

Although the lower mountain slopes have their exclusive residents, it is on the higher reaches, from 6,000–10,000 ft above sea-level up towards the summit, that wildlife shows the greatest adaptations to mountain life. The harsh climate, the lack of oxygen and the steepness of the terrain have steered the course of evolution for animals of the high mountains.

Animals combat cold temperatures, particularly at night, through a combination of physical adaptations and modifications to their behavior. A notable feature of many mountain species that have lowland relatives is their relatively large size. Large animals lose heat more slowly than smaller creatures, so there is a distinct advantage in having body bulk at high altitude. Pumas range the length of the Americas, where they inhabit a variety of habitats, from the tropical rain forests to the mountain heights. The highland races of the puma are larger and

ABOVE The chinchilla thrives at altitudes approaching 20,000 ft in the South American Andes. If often feeds after dark on grass, mosses and lichens. Most animals of its size would rapidly lose their body heat in the freezing night-time temperatures at such extreme altitudes. But the chinchilla's dense, silky fur provides it with remarkable insulation.

Backs To The Blizzards

The cold, wind and snow of the high Tibetan plateau make the region one of the bleakest on earth. Few animals can survive in the harsh environment, but the yak – one of the 12 species of wild cattle – is an exception. It lives entirely at high altitudes. In winter, it dwells at over 15,000 ft on the plateau and in the valleys, and in summer, it forages on the valleys, and in summer, it forages on the permanent snowfields of the peaks, moving up as high as 20,000 ft.

Double insulation

The yak is a massive creature that weighs up to a ton and measures over ten feet in length. As an adaptation to its remote habitat, it has evolved a warm coat that provides a double layer of insulation. The thick, woolly underfur is covered by a long, coarse winter coat that drapes over its hindquarters, legs and forehead. Even its nose is hairy (except for a thin band of skin on it's upper lip). In summer, when daytime temperatures are warm, the yak molts patches of its winter coat to prevent it from overheating. During bitterly-cold blizzards, which regularly occur even in June, the yak turns its well-protected rump into the wind and waits for the snowfall to cease.

Yaks rarely drink, obtaining most of the moisture they need by eating snow. They graze on the sparse vegetation – mainly grass and lichens – and keep their activity to a minimum as a means of conserving precious energy.

RIGHT The yak is perfectly at home at altitudes of over 15,000 ft – only the giant peaks such as Nuptse in Nepal are beyond its reach. The yak has a double coat of fur to ward off the blizzards. it eats sparse plant life such as lichens and swallows snow to gain water.

almost twice as heavy as their lowland counterparts. Though most hummingbirds are so tiny that they cannot keep warm in cold conditions, one Andean species, the giant hummingbird, has turned back from miniaturization. At 8 in in length, it is about the same size as a swallow – its larger size playing a part in reducing heat loss at altitudes of 12,000 ft or more.

A common evolutionary trend among mountain animals is the growth of a thick coat for insulation against the cold. Dense, shaggy fur is a feature of many hoofed mammals from the uplands, including the Himalayan tahr and the markhor form Asia, and the mountain tapir and the vicuna of the Andes. Both the Tibetan yak and the Rocky Mountain goat have double-layered coats that protect them from the cold at extreme heights (see boxes on pages 322 and 342 respectively). For the rodents, small body size makes insulation essential. The chinchilla and the mountain viscacha of the Andes have such dense, silky fur that both have been widely hunted for their pelts. No fewer than 60 hairs can sprout from the same follicle in the chinchilla's skin, providing such an efficient barrier against the cold that the animal can survive at altitudes of 20,000 ft.

Burrow sharing

Few mountain animals are nocturnal, since they cannot function well in the freezing night-time temperatures. Most keep night-time activity to a minimum to reduce the expenditure of energy. As they breathe, the animals's long, narrow nostrils warm the night air before it enters their lungs. Many animals shelter from the wind in caves or among boulders. Small mammals, such as marmots and pikas, dig extensive burrows in alpine meadows. Insulated from extreme changes in temperature by the layers of soil and rock above, the burrows remain sufficiently warm for the animals to rest there after sunset. Many birds, including hawks, pipits and flycatchers, roost and nest in cracks and holes among rocks; some even share burrows excavated by rodents. Alpine marmots may find their homes commandeered by little owls, while in the Himalayas, pikas may spend the night in the company of snow finches.

Groups of roosting birds sometimes spend the night in more intimate fashion, huddling close together so that each bird reduces the surface area it exposes to the cold. The mutual warmth provided by communal roosting is often so effective that over 200 birds may occupy a single rock cleft.

ABOVE Pikas, such as the collared pika of North America, protect themselves from the night-time cold by sheltering among rocks or by digging burrows in the high-altitude meadows. The rock and soil above their shelters insulates them from the sharp changes in temperature that occur on the surface. Burrowing animals sometimes provide shelter for other creatures, and pikas that live in the Himalayas often share their burrows with snow finches.

Despite their small size, several species of hummingbirds (besides the giant hummingbird) inhabit high mountains. Species such as the Andean hillstar and the broadtailed hummingbird of the Rockies survive the chill of night by greatly slowing down their metabolisms until they reach a state of torpor.

ABOVE The vicuna grazes the high plateaux of the Andes, where its thick fur provides vital protection from the cold nights. In the daytime, however, the dense hairs prevent the vicuna losing excess body heat in the strong tropical sunshine. To regulate its temperature, the animal stands with its legs apart, enabling the air to flow around the inside of its thighs, where the hair is especially thin and heat can readily escape.

Like hibernating mammals, they move into shelter and reduce their body processes to an absolute minimum, enabling their internal temperatures to drop by as much as 12 F. Torpidity is extremely rare in birds, but the tiny size and high energy expenditure of hummingbirds makes the tactic essential.

Soaking up the sunshine

In the daytime, mountain animals put the warmth of the sunshine to good use. The few lizards that inhabit the upland zones gather on bare rock, for a few hours each day to bask in the sun and raise their temperatures to a working level. Many of the insects that live at high altitude, including flies, beetles and butterflies, have dark colors that absorb as much as possible of the sun's heat. Some harlequin frogs of the northern Andes have jet black backs to soak up the equatorial sun. Dark pigments also provide screening from ultra-violet radiation.

Mountain bird species build their nests in the warmest available places to ensure that their eggs do not chill. They usually locate nests in shelter from the wind and, if possible, in a position where they can receive the rays of the daytime sun. In the Tien Shan, the Himalayan rubythroat prefers to breed on south-facing or eastfacing slopes where the sun shines directly onto the nest. At midday, especially in the tropics, the sunshine can become so strong that the eggs may be in danger of overheating. Some birds site their nests under rocks or clumps of vegetation in such a way that they receive direct warmth in the morning, when the sun is low in the sky, but fall under shade during midday, when the sun is high and its rays intense.

With their luxuriant fur, some mountain mammals are also in danger of overheating during the day, especially if they expend energy when running or fighting. On the inside of their thighs, vicunas (South American animals related to llamas) have patches of sparse hair that act as heat regulators. When they are

cold, vicunas stand with their legs together so that the bald patches are not exposed to the air. When they need to lose heat, they keep their legs apart, allowing air to circulate around their thighs and draw heat from their blood.

Fierce winds

The fierce winds that blow across the upper slopes of mountains not only chill the land, but also make flight a hazardous undertaking. Birds cope with the gales in one of two ways. Some, particularly among the birds of prey, are especially large, with powerful wings that can keep flight stable in the buffeting winds. Some of the largest of the world's vultures inhabit the peaks; they include the Andean condor, the lammergeier, the griffon vulture and the black vulture. The birds use the swirling winds to soar over the slopes in search of carrion. Smaller birds, such as pipits and wheatears, avoid the wind by keeping close to the ground and sheltering behind rocks. When they take to the air, they usually fly low over the surface where the wind is often less fierce.

Winter whitener

On the slopes of the North American and Eurasian mountains, cold, inhospitable conditions are intensified in winter. Temperatures drop dramatically, snow accumulates well below the permanent snowline, and the wind sends blizzards to whiten the sky. Vegetation shrinks back and disappears under a blanket of snow, and avalanches thunder down the slopes, sweeping away in their path.

ABOVE The long, broad wings of large birds of prey, such as the lammergeier, enable them to overcome the buffeting winds on the mountaintops. One of the largest of the vultures, the lammergeier soars over mountains in Europe, Asia and Africa, searching for carrion. Especially fond of bone marrow, the lammergeier will drop bones onto rock to break them open.

Life becomes extremely difficult on the mountaintops and only a few animals are active in the snow. The ptarmigan is one of the most hardy mountain birds and remains in the alpine zone through all but the worst blizzards. It molts in the winter, replacing its mottled brown summer plumage with feathers of pure white that act as perfect camouflage in the winter snows. The mountain hare similarly molts its coat in winter, exchanging its

LEFT The white winter fur of the mountain hare changes color to blue-brown when the snow cover recedes in spring. The mountain hare lives above the treeline on meadows and moors, where it grazes on grass and low shrubs such as heather. It has many enemies on the mountains, including wild cats and golden eagles. The adults rely on their rapid sprinting speed for escape, but the younger, more vulnerable hares take refuge in burrows in the turf.

brown fur for a bluish-white coat. Ibex move onto steep, rocky slopes that gain plenty of sunshine. Deep snow cannot accumulate on the crags, and vegetation is easier to find.

Safe beneath the snow

Most animals take steps to avoid the worst of the winter conditions. Insects bury themselves in the soil and hibernate, protected from the wind and sudden drops in temperature by the blanket of snow that covers the ground. Burrowing rodents hibernate in their underground shelters. Alpine marmots retire deep into their burrows – as much as ten feet from the surface – and enter a prolonged sleep that may last more than eight months. Pikas rest in burrows or rock crevices, relying on stores of food collected in the preceding months to see them through the winter while snow covers the ground vegetation.

Escape downhill

During the winter, many birds leave the mountains and fly to more southerly habitats before the chill sets in. The yellow-rumped warbler leaves the Rockies for the warmth of California and Mexico. Himalayan rubythroats stay in the Tien Shan range for only a short time. They arrive in the spring from India, nest and rear their young, and then return south as early as August.

Shelter under the trees

Migration does not always entail a marathon journey. Many mountain animals make use of the changing climatic conditions downslope to escape the worst of the winter weather. By trekking or flying a mile or so downwards, they encounter a rapidly changing environment. Sometimes, shaded valleys become traps for sinking, cold air, and forested slopes accumulate snow (from snowfall and from avalanches) but the lower slopes generally provide a more comfortable winter habitat. Chamois descend from the alpine zone to the conifer forests in the mountains of Europe, while markhor and bighorn sheep cross below the treeline in central Asia and the Rockies respectively. They do not have to compete with the hoofed mammals that normally range up to the treeline, because they too move downslope. In the Alps, red deer find the conifer belt too harsh in winter and migrate down to the deciduous woodland below.

Birds also benefit from a downwards migration. Wallcreepers fly to warmer valleys from the rocky crags of the Alps, and river chats fly downhill in the Himalayas. In the Southern Alps of New Zealand, the small surviving population of the flightless takahes walk downslope. In summer, they forage and breed on the grasses of damp, alpine meadows up to 6,000 ft above sea-level. When winter snow buries the grass, they trek down to valleys cloaked in southern beech trees, where ferns and herbs provide an alternative food source.

ABOVE The takahe is one of many mountain animals that live at different altitudes during different seasons. A flightless bird that inhabits the mountainous Fiordland region of New Zealand, the takahe breeds in the alpine meadows up to 6,000 ft above sea level. When winter brings snow to the mountains and buries its food supply, the takahe must take evasive action. By walking downhill into the forested valleys, it escapes the worst of the winter weather.

Altitude sickness

On the mountaintops the air is thin and contains little oxygen, creating severe problems for the wildlife. When humans and other vertebrates breathe in, air enters the lungs and oxygen from the air passes into a network of blood capillaries beneath the surface of the lungs. Oxygen travels around the bloodstream in the red blood cells that give the fluid its color. The cells contain haemoglobin – a substance that combines chemically with oxygen. As the blood circulates through the body, oxygen steadily diffuses from the blood vessels into the cells of organs, muscles and other tissues. Oxygen is essential for respiration – the chemical process that provides energy for living cells.

Animals have to adapt to the shortage of oxygen at high altitude in order to survive. Above 10,000 ft, it becomes increasingly difficult to breathe, and in humans, fatigue sets in. Too rapid an ascent or excessive exercise can lead to altitude sickness. Symptoms include vomiting, headaches and breathlessness – all caused because insufficient oxygen is entering the bloodstream and reaching the bodies' tissues. If untreated, altitude sickness can be fatal. One effect is the redistribution of body water, causing the lungs to flood with fluid. Meanwhile, deprived of oxygen, brain cells start to die.

A more gentle exposure to high altitude gives the body time to adjust to the lower oxygen supply – a process called acclimatization – and many people who live permanently in mountains, such as the Andean Indians, are dependent on it. Acclimatization improves the passage of oxygen into the bloodstream, largely by increasing the number of red blood cells circulating around the body.

Animals that live at high altitude possess an even greater concentration of red blood cells, combined with a more efficient take-up and release of oxygen by the haemoglobin. Vicunas from the high Andes have 14 million red blood cells per cubic millimeter, compared with a maximum of only six million in humans. Every time vicunas breathe in, a much greater proportion of oxygen passes from the thin air into their bloodstream. Efficient oxygen diffusion around the bodies of bar-headed geese enables them to fly at incredibly high altitudes. The geese breed on damp meadows high in the Tibetan plateau, and migrate south in winter across the vast wall of the Himalayas. They even occur in the sky above Everest.

Lake Titicaca

Aquatic animals require oxygen and, at high altitude, there is a corresponding shortage of oxygen in water. Lake Titicaca is located in the Andes, 12,500 ft above sea-level. One of its inhabitants, the Lake

ABOVE Sheer slopes, such as those in the Navajo National Monument park in the USA, offer few opportunities for wildlife. Plant life cannot take hold on the rock-face and there is little food in the crevices to support animals. But such slopes can provide vital refuge and shelter for some mountain animals. Nimble goat antelopes can pick their way across the cliffs to escape predators, and the rock-faces offer secure nesting sites for many birds.

Titicaca frog, has overcome the problem of oxygen shortage. It has a greatly increased surface area in the form of baggy skin that hangs from its back, sides and hindlegs. A dense network of capillaries fills the folds, and the rich blood supply absorbs as much oxygen as possible from the water through the extra skin.

Cliffhangers

High on the mountains, where steep precipices and frost-shattered rock reach up to the summits, movement becomes precarious for the animals. In order to cope with the steep slopes and rocky terrain, the animals must be agile and sure-footed. Several of the goat antelopes show extraordinary climbing ability among the crags. Ibex habitually make their way across seemingly impassable cliff faces to perch on rock ledges thousands of feet above a valley floor. Chamois are so nimble that rivals dart around the crags in pursuit of one another during the rutting season, with little danger of a slip or a fall.

The goat antelopes and llamas of the Andes have specially structured hooves with flexible pads on their soles that provide tread in the same way as a car tire grips the road. The pad of the Rocky Mountain goat is surrounded by a hard, projecting rim around its hoof that catches on ridges and irregularities on the surface of rocks. Crags and cliffs in the low mountains of Australia are the home of rock wallabies – relatives of the lowland kangaroos – that climb and spring among the rocks with such agility that they can leap chasms 12 feet wide. They have treaded pads on their hindfeet, stiff hairs around the soles and short claws that provide a grip on the stones.

Some smaller animals live entirely among the rocks, resting in fissures and finding all the food they need among the cracks and hollows. In New Zealand, the rock wren seldom leaves the alpine zone. It remains out of sight for most of the time, searching for insects among the stones. It is active even in winter when snow and chill winds make foraging in exposed sites impossible. Many other birds nest in hollows in the rocks, and some species will only site their nests on steep rock faces that are inaccessible to terrestrail predators. Golden eagles and Verreaux's eagles balance their bulky nests on cliff ledges, while the crag martin forms its nest from pellets of mud stuck onto smooth, vertical rock faces.

Life on top of the world

The glaciers and snow that cover the highest mountain peaks represent the mountain environment at its most extreme. Birds of prey occasionally soar around the snow-fields and migrating birds fly through the skies above the summits, but they are brief visitors. The icy mountaintops of the Himalayas and the high Andes cannot support large animals, but invertebrates exist in large numbers.

A community of tiny creatures lives among the snowfields. Their food chain consists of pollen, seeds, the bodies of insects and other organic debris that sweeps up on the wind from lower altitudes. Springtails, birstletails, anthymiid flies and grylloblattid insects feed on the debris. As they feed, they become the prey of salticid jumping spiders that hunt over the snow during the daytime. Mountaineers on Everest have recorded such spiders at more than 20,000 ft above sea-level.

Anti-freeze

Many of the high mountain invertebrates have anti-freezing substances that lower the temperature at which their body fluids solidify. They can cope with a prolonged period of development, and in the cold, their body processes operate so slowly that their life-cycles take years to complete. Five years may elapse before larvae muster the resources to turn into adults. The great majority of high-altitude insects are wingless or have tiny, non-functioning wings. In the high winds, crawling is more effective than attempts at flying, and, in any case, the extreme cold would make the vigorous muscle action necessary for flight impossible. To grow wings would be wasteful of the precious energy that the insects glean from their mountaintop environment.

ABOVE RIGHT Rock ledges on sheer crags are the favorite nesting sites for many birds of prey. Here, a Verreaux's eagle rears its young on a rock-face in the Matapo Hills of Zimbabwe. In order to avoid predators, it nests in the most inaccessible places on cliffs, constructing a platform of sticks that often measures over six feet across.
RIGHT Common in the Alps as well as along rocky shorelines, the crag martin forms breeding colonies in the spring. Each pair builds a nest from pellets of mud that they plaster against the rock wall. They often position the nest beneath an overhang where they can shelter from the rain and the midday sunshine.

ABOVE The red panda inhabits the rhododendron groves and coniferous forests on the lower slopes of the Himalayan range. Mainly active at dusk and at night, it rests by day curled in the fork of a branch. Though it is an agile tree-climber, with hairs on its soles that improve its grip, the red panda frequently comes to the forest floor to feed. It eats a variety of food including bamboo leaves and shoots, fruit, insects, eggs and young birds.

The mighty Himalayas

The Himalayas, the loftiest of all the world's mountain ranges, offer wildlife a variety of habitats, including some of the most testing conditions on earth above the treeline. The range forms a great arc that sweeps from Kashmir through Nepal into northern Burma. To the south, the uplands begin with the forested slopes of the foothills. From there, the land rises abruptly until it reaches a chain of mighty peaks, 25,000 ft or more above sea-level. To the north, however, the land falls away only a few thousand feet before the Himalayan mountains grade into the vast Tibetan plateau, almost all of which lies at over 12,000 ft.

The jagged peaks that crown the Himalayas form a collection of mountains without rival among the other ranges of the world. Scores of peaks along the border of China and Nepal soar to a height far in excess of the tallest mountains of Europe, Africa and the New World. Four out of five of the world's highest mountains occur in the same region – Everest (29,000 ft), Kangchenjunga (28,000 ft), Makalu (27,800 ft) and Dhaulagiri (26,800 ft). The ice-capped giants, particularly those in the eastern Himalayas, receive plentiful snow. The moisture-laden monsoon winds that blow across India, deposit their last reserves of water as they rise over the mountain wall. They bring heavy rain to the foothills and deep snow higher up.

Among the rhododendrons

North from the plain of the Ganges and the valley of the Brahmaputra, the Himalayan foothills start to rise, with rich, humid forest cloaking their lower slopes. Many creatures native to the monsoon forests of India roam through the habitat, including gaur (wild oxen) and their chief enemy, the tiger. But above 3,000 ft the character of the forest changes and rhododendrons dominate the slopes.

Growing to a height of some 30 ft the rhododendrons produce a dense mass of glossy leaves and springtime blossoms of red, pink and white. The leaves, which contain a high proportion of silica, are unpalatable to many herbivorous animals, but the flowers provide a rich source of food. Insects crawl over the blooms, and sunbirds dip their beaks inside to sip nectar, transferring pollen as they come and go. Himalayan races of the Hanuman langur, with slightly thicker coats than their lowland cousins, are not so beneficial to the flowers in their feeding habits. The monkeys simply pluck handfuls of blossom from the branches. Other creatures exploit the concentration of insects that gathers around the blooms. The fire-tailed myzornis is expert at snatching insects from the blossoms using its slender bill. A small, bright green babbler, with red tail-feathers, it is a characteristic feature of the rhododendron belt, seldom foraging at lower levels.

The Himalayas are rich in gamebirds, particularly pheasants and their relatives, and some of the most colorful species inhabit

ABOVE Several colorful gamebirds inhabit the Himalayas, ranging from the mountain forests to the meadows above the treeline. The Himalayan monal is a true mountaineer, reaching heights of up to 15,000 ft above sea level, only retreating below the treeline in winter if there are no snow-free pastures above. It uses its strong beak to dig into the soil for roots and bulbs, sometimes excavating holes 1 ft in depth.

the rhododendron groves. Western tragopans and satyr tragopans nest in the branches but feed on the forest floor, picking through the leaf litter for insects, fresh leaves, buds, seeds and berries. The intricate, speckled plumage of the male western tragopan erupts into a blaze of color when the bird displays to a female. He erects two long, blue, fleshy horns on his head, and inflates a broad pouch of bare skin on his chest that is strikingly patterned with pink, purple and pale blue.

Conifer dwellers

In many parts of the Himalayas, the rhododendron thickets thin out at about 8,000 ft to be replaced by conifers, such as Himalayan fir. These trees are more resistant to the increasing cold, and their thin, needle-shaped leaves easily shed the snow that regularly falls. Some of the wildlife from the rhododendron belt also occurs in the conifers, including the tragopans. Other animals, such as the Asiatic black bear and Sclater's monal (a type of pheasant), forage both in the conifers and in the meadows above. Some creatures, however, are adapted to live only in the coniferous forests. The Himalayan race of the common crossbill rarely occurs far from the pine trees in which it nests and feeds. Its bill has a special cross-tip design enabling it to clip seeds from the pine cones.

The conifers are the stronghold of the red panda – a smaller, slimmer relative of the giant panda. Spending most of the time in the branches, the red panda searches at night for a range of plant and animal foods, including the occasional small vertebrate. It is an agile climber with a fuzz of hairs on the soles of its feet that provide a firm grip on branches. Its thick, red-brown fur keeps the chill at bay, and blends surprisingly well with the color of the tree bark as it rests during the day.

Above 10,000 ft, the conifers thin out and are replaced by an open landscape of tussock grasses, shrubs,

ABOVE The Himalayan tahr haunts rocky, wooded slopes in the Himalayas. Agile and sure-footed, with a dense, warm coat, it is well equipped for the mountain environment. Male Himalayan tahrs tend to stay on the forested slopes the year round, but the females move up to the mountain pastures in summer, where they gather in groups of several animals. They sometimes forage in company with Himalayan monals.

cushion plants, alpine flowers and rocks. As the snowline approaches, the vegetation becomes more and more scant, while beyond it, few animals – other than the invertebrates that feed on wind-blown debris – can survive. Between the treeline and the snowline, however, live a variety of hardy animals. Marmots and pikas dart among the rocks, nibbling the plants; Royle's pika, for example, can find enough to eat at heights of more than 15,000 ft. Birds, such as Himalayan accentors and the grandala, nest in the alpine meadows, but fly to warmer elevations for the winter. The long, pointed wings of the grandala give the bird fast, controlled flight in spite of the strong winds that roar across the alpine zone.

The Himalayan monal spends most of the time walking among the boulders of the alpine zone, only taking to the air when in danger. When it does so, it puts the steep, open nature of the terrain to good use. After a few flaps to get itself airborne, the Himalayan monal glides downhill, using gravity to ferry it hundreds of feet from danger in a matter of seconds. No such tactic is available to the Himalayan tahr, but the animal has the benefit of agility and a firm hoofhold on steep, rocky terrain. Large size – the Himalayan tahr may stand 3 feet at the shoulder – and a thick coat enable the species to retain heat high in the mountains.

A thick pelt also insulates the tahrs greatest enemy, the powerful snow leopard. An exclusive resident of the mountains, the snow leopard hunts throughout the peaks of central Asia. It inhabits alpine meadows and crags up to 15,000 ft high, only retreating towards the treeline in the depths of winter. Against, the gray rock and the snow, its pale coat provides perfect camouflage, allowing it to ambush goat antelopes, marmots and birds. The cat prepares its den in a crack or cave in the rock, sometimes lining a crack with hairs from its coat to provide the cubs with the extra protection from the cold.

Summits of Europe

Two mountain ranges with similar wildlife – the Alps and the Caucasus – contain the highest peaks in Europe. The Alps form a broad curve that begins near the Mediterranean coast along the border of France and Italy, and sweeps through Switzerland and Austria as far as northern Yugoslavia. The mountains are at their most dramatic in Switerland, where broad, deep valleys lie thousands of feet below jagged, ice-capped peaks. Several summits rise above 12,000 ft; they include Monte Rosa (15,200 ft) and its neighbor, the Matterhorn (14,700 ft), both of which straddle the Swiss-Italian border, as well as Mont Blanc (15,780 ft), the highest mountain in the Alps, which bestrides the frontier of France and Italy.

1,200 miles to the east of the Alps, on the fringes of Europe, stand the Caucasus. The range forms a striking narrow line of mountains

that runs from the Black Sea to the Caspian Sea, entirely within the boundaries of the USSR. In only a few places is the range wider than 100 mi. Along the southern edge, the Caucasus rise to heights of more than 13,000 ft, but on the northern side, the mountains slope more gently. The highest peak is El'brus (18,500 ft), followed by Dykh Tau (17,000 ft) and Koshtantau (16,870 ft).

Changed and unchanged

The natural vegetation of the Alps changes with the increase in altitude – from deciduous forests to conifers and alpine meadows. A long history of settlement, however, has had a profound effect on the landscape, and much of the vegetation below the 7,000 ft treeline has now been cleared to make way for villages and fields. In the Caucasus, much of the original forested land remains, showing how the Alps must have looked centuries ago. Deciduous forests of oak and elm, with beech and then birch and maple on higher ground, eventually give way to dark forests of spruce. In many places, particularly in the west where the rainfall is highest, rhododendrons occur close to the treeline. Beyond the treeline are colorful, sub-alpine meadows, with grasses and herbs growing over 6 feet tall, while even further up the slopes are the alpine meadows with their low grasses and cushion plants. The mountain summits of both the Alps and the Caucasus harbor glaciers and permanent snowfields.

Last refuges

Many of the animals that live below the treeline in the Alps and the Caucasus reflect the wildlife of the woodlands and forests further north in Europe. Black, woodpeckers and common crossbills, for example, inhabit the coniferous forests in both mountain and lowland habitats. Some animals that are closely associated with mountain forests are there simply because they have been driven out of their native habitats. The surviving trees that remain undisturbed on the steep mountainsides provide them with a last refuge.

ABOVE The snow leopard is a rare and elusive predator of the Himalayas and other central Asian mountains. It hunts high above the treeline in summer, ambushing birds and mammals up to the size of tahrs. Its thick fur keeps out the cold, and the pale coloration of the hair provides it with excellent camouflage against the rocks and the snow.

PAGE 334 Beech woods are common on the lower slopes of European mountains, creating shady woodland with little ground vegetation. Many lowland creatures roam onto the wooded lower slopes, including deer, wild boar and lynxes.

GLOSSY OFFSPRING

The alpine salamander inhabits the Alps and the mountains along the east of the Adriatic Sea. Restricted to mountain zones, it seldom appears below 2,200 ft and frequently ranges to heights of 10,000 ft in the high meadows well above the treeline. Measuring about 6 in in length, the alpine salamander has glossy, black skin. A secretive animal, mainly active at night, it searches the meadows for earthworms, slugs and insects. By day, it hides in a moist spot, usually under rocks, among vegetation or beside waterfalls where the spray dampens the ground. It avoids the cold winter weather by moving into cover and hibernating for several months.

The alpine salamander's most extraordinary adaptation to its environment takes place during its earliest stages of development. Most salamanders lay their eggs in pools, and the aquatic larvae that hatch continue their growth underwater. However, ponds on the high slopes regularly freeze, and remain frozen well into the summer. To overcome this problem, female alpine salamanders retain the eggs inside their bodies, where they hatch. The larvae continue to grow inside the females, and develop extremely large red gills that are able to absorb the nutritious fluid surrounding them. At the points where they press against the mother's uterus, the gills absorb oxygen directly from her tissues. In the cold, alpine temperatures, the young take at least two years to develop, and the female finally gives birth on land to two fully-formed salamanders. At only two inches in length, they are miniature versions of their parents, but they are immediately ready for life on the mountain slopes.

BELOW Few amphibians range as high above sea level as the Alpine salamander, which thrives above the treeline in the Alps. Hiding under rocks or among vegetation by day, it hunts over the grass at night, in search of insects, worms and slugs. it seizes its prey with a sudden lunge and swallows the victims with violent shakes of its head.

Brown bears still roam the forests of the Caucasus, gaining most of their food from the plants on the forest floor, but occasionally trapping and eating small animals. In the warmest months, they emerge to forage on the meadows above the treeline. The same forests also provide a refuge for wolves and lynxes. Red deer, roe deer and wild boar provide food for the large carnivores. Hunting and habitat loss have long since pushed most carnivorous mammals from the Alps, although small species, such as the wild cat and the beech marten, survive.

High-flying Apollo

Some of the most distinctive and specialized wildlife communities occur on the higher slopes of the mountains. The Apollo butterfly, which visits the alpine flowers, is a true mountain-dweller and seldom appears outside the open, upland landscape. The adults are active only on the warmest days of the year, laying their eggs on plants such as stonecrop that grow among the boulders. The bag-worm moth lives higher still, in some cases close to the glaciers; the females are flightless and dwell under rocks.

The high slopes of both the Alps and Caucasus are too cold for most reptiles and amphibians, but there are a few exceptions. The viviparous lizard survives at up to 10,000 ft in the Alps, while the alpine salamander is entirely confined to mountain zones (see box on page 335). Small mammals, especially rodents, are more numerous than reptiles, though they are still few in comparison with the numbers of mammals that exist in warmer habitats.

ABOVE The Alpine chough seldom strays far from mountains, and is widespread in the heights of the Alps. Though it moves down the slopes in winter, it rarely descends below 5,000 ft. In summer, the bird dwells well above the treeline, often walking across high-altitude meadows to look for insects and spiders. It also lingers near tourist huts and mountain restaurants, ready to seize scraps of food.

Most of the mountain rodents find shelter in the open terrain by burrowing, but some take subterranean living to extremes. The long-clawed mole-vole of the Caucasus spends almost all its time underground in the high-altitude meadows. Feeding mainly on plant roots, it digs a series of burrows radiating from a central nest in its search for food. The alpine marmot, on the other hand, forages for plant food on the surface, only using its burrow for shelter and hibernation. A colonial species, it builds complex underground havens that have a network of tunnels and chambers. Alpine marmots are always on the alert for danger, standing erect on their hindquarters to obtain a better view of their surroundings. As soon as a predator, such as an eagle, appears overhead, the look-out gives a shrill, warning cry that sends all the members of the colony – and even neighboring colonies – racing underground.

ABOVE The chamois is one of the most nimble of all mountain dwellers. When in danger, it flees onto sheer rock faces, often leaping across gaps of several yards to escape its predators. Pseudo-claws on the back of the chamois' feet provide extra grip on steep, grassy slopes. When facing downhill, the chamois presses its hooves deep into the turf so that the pseudo-claws press into the ground.

Mountain birds

The vertebrates that are most successful in coping with a high-altitude life are birds. Numerous species breed in the alpine zone, from seed-eaters and insectivores to the carnivores and scavengers of the high mountains. The snow finch remains on the alpine meadows for most of the year, only retreating beneath the treeline when the worst of the winter weather arrives. The alpine accentor follows the same pattern, but the rock thrush is a long distance migrant that leaves the mountains to fly to Africa for the winter. All three species are common in the Alps and the Caucasus, building their nests among the rocks and boulders or within hollows in the ground.

The wallcreeper is one of the most colorful of upland species, flashing its bright red wing patches as it flutters from one cliff face to the next. An agile climber, the wallcreeper forages on the crags, poking its long, downcurved bill inside rock crevices to catch insects. Like the wallcreeper, the alpine swift nests in cracks in the rock faces, but finds its food in the sky. With a wingspan of nearly 2 ft, the alpine swift has a flight that is sufficiently powerful to combat the fierce winds that blow over the peaks, allowing it to race through the skies and snap up flying insects. When feeding its young, the alpine swift clocks up hundreds of miles every day as it flies in pursuit of food and takes it back to the nest.

Gamebirds are not as diverse in the European mountains as they are in the Himalayas, but several species do occur. Ptarmigan live high in the Alps, while the Caucasian snowcock inhabits only the Caucasus. Ascending to 12,000 ft above sea-level in summer, the snowcock moves down to warm, south-facing slopes in the winter. There are five species of snowcocks in mountains throughout Eurasia, and all have similar habits. They feed on plant food, particularly bulbs, and will often take advantage of holes that goat antelopes have dug in the snow to reach the vegetation beneath.

Soaring scavengers

Ravens are members of the crow family and occur in many wild places, although they are particularly common in mountains, where they make their home on lonely crags. Soaring on broad, outstretched wings, they search the slopes for carrion, but will kill a variety of small animals with heavy blows from their beaks. The alpine chough, a species of crow, is even more characteristic of the mountains, and occurs in few other habitats. Using the high winds, it soars on the updraughts around the cliffs, where colonies of choughs build their nests.

Some of Europe's most spectacular birds of prey inhabit the high mountains. The golden eagle hunts for a variety of animals, ranging from marmots and game-birds to young goats. In the Alps and in many parts of the Caucasus, the golden eagle is the most powerful of all the predators. An even larger bird of prey haunts the Caucasus – the lammergeier or bearded vulture. A huge creature well over 3 ft in length, the lammergeier is primarily a scavenger, but it kills small mammals on occasions. To extract is victim's bone marrow, the lammergeier drops the bones from a great height onto the rocks below. When they shatter, the bird swoops down to pick at the exposed marrow.

Ibex on the rocks

Goat antelopes, such as ibex and chamois, are the largest animals to frequent the high slopes of the Alps and the Caucasus. Ibex climb the steepest rocky pinnacles, and rarely move down to the valleys, even in winter. The male ibex is much the larger of the two sexes, weighing more than 320 lbs. It grows a pair of stout, curved horns that it interlocks with those of rivals during ritual combat. The chamois is more lightly built and picks its way nimbly among the boulders. Small herds of females and their young often file across the high meadows in search of better pasture; the males remain solitary, except during the rutting season.

Caucasian turs

Both ibex and chamois forage for leaves, roots, mosses and lichens, often clearing away light snow to reach their food. In addition to the chamois and the ibex, two species of goat antelopes – the East Caucasian tur and the West Caucasian tur – and the wild goat inhabit the Caucasus. While the turs keep to the high meadows, the wild goat – the probable ancestor of the domestic goat – only ventures to the upper slopes from time to time. It usually keeps to rocky terrain from 3,000–8,000 ft above sea-level.

Continental chains

The great mountain ranges of the Americas stand apart from those in the rest of the world, not merely for their vast length, but also for their orientation. Instead of crossing from east to west, like most of the Eurasian mountain ranges, they wind from north to south, taking in both hemispheres. From the snowbound peaks of Alaska on the fringes of the Arctic, through the broadening, parallel ranges of the Rockies and the Coast Ranges, the mountain chain runs all but continuously south to the tropics. Narrowing through the land bridge of Central America, the mountains emerge again above the lush rain forests of Amazonia to form the Andes. Running for over 5,000 mi as they trace the western coast of South America, the Andes are the longest single range in the world. Crossing the Equator in Ecuador, they stretch further south, forming a narrow but intact wall which no river or valley crosses. Continuing past the Tropic of Capricorn, they eventually reach the icy, storm-wracked tip of Tierra del Fuego in the sub-Antarctic.

The combination of altitude and changing latitude create a great variation in habitats along the length of the mountain chain. On the lower slopes of an Alaskan mountain there may be tundra or even permanent snow, while further south, the habitat could be coniferous forest, temperate woodland, desert or luxuriant tropical forest. One effect of the north-south orientation is that vegetation normally specific to one area extends its distribution along the mountain chain. In the cool mountains, tundra plants and conifers that are adapted to cold climates, extend much closer towards the tropics than they would otherwise range.

Some animal species, such as the puma, roam almost the entire length of both continents, but generally the mountain fauna changes dramatically from the Rockies and mountains of tropical America to the southern Andes. Despite the similarities in climate, the animal life at opposite ends of the mountain chain is quite different.

Wintry Rockies

The mountains of western North America form a complex system of ranges, dominated by the Rockies and the coastal ranges of Canada and the USA. They reach their highest point at Mount McKinley (20,230 ft) in Alaska, where extensive glaciers scour the heights, and spread south with lower summits through British Columbia into western USA. As they progress south, the snowline and the treeline steadily climb in tandem with the warming of the climate. By the time the ranges reach south-western USA, hot deserts skirt their feet. The highest peak in the mountains of Canada is Mount Logan (19,850 ft) in the Yukon Territory, while in the Sierra Nevada, Mount Whitney (14,495 ft) stands as the tallest summit in mainland USA.

Winter in the northern mountains is bitterly cold and the ravages of the climate have a profound imapct on wildlife. Most of the larger animals move downslope, either below the treeline or, in the case

of the wapiti, down into the sheltered valleys. Many birds from the high altitude meadows migrate away from the mountains. Townsend's solitaire is a thrush that ranges from Alaska to California, breeding in the upland forests and in the open country above. Individuals from the northern part of the range fly south when winter sets in. The birds that breed in the south remain there all year round and are able to find enough berries in the mountains to sustain them through the winter.

The tailed frog of the Rockies and the Cascades shows a variety of adaptations that enable it to cope with the rushing, turbulent waters of mountain streams. The tailed frog takes its name from a tubular organ with which the male fertilizes the female internally. The structure is unique among frogs, and avoids the risk of sperm sweeping downstream before fertilization has taken place.

Some animals that inhabit the north American Rockies extend their range into the conifer forests on

ABOVE At 20,230 ft, Mount McKinley, in the heart of Alasks, is the highest summit in North America. Standing on the fringes of the Arctic, it has a cold climate from top to bottom. Snowfields and glaciers extend for several thousand feet down its slopes. The mountain stands at the top of an enormous chain of peaks that run through North, Central and South America. Progressing south from Alaska along the chain, the snowline steadily increase in altitude as the climate improves. In the equatorial zone of South America, permanent snow merely forms a cap on peaks that rival Mount McKinley in height.

ABOVE The bighorn inhabits alpine meadows and rugged foothills in western North America. More at home in rocky terrain than other wild sheep, it climbs steep slopes and darts among the rocks when threatened by predators such as pumas. Spectacular battles occur between rival male bighorns during the breeding season. The adversaries rear up on their hindlegs, lunge forwards with their heads lowered, and crash their massive, curled horns together with a thud that echoes around the crags.

more southerly mountains. Grizzly bears survive in some mountain reserves in the USA, although they are much more common further north in Canada and Alaska. Most of the bear population in the USA was exterminated by settlers who feared for their lives and their livestock. But grizzly bears are not primarily carnivores. Like other brown bears, they usually feed on berries and shoots, supplementing their diet with some small animals, and they only attack large mammals on rare occasions.

Hunters on the mountain slopes include golden eagles that roam the Rockies (as well as the peaks of Eurasia) and pumas (more commonly known as cougars). Both predators hunt above the treeline, where they are the chief enemies of goat antelopes. The golden eagle snatches only the lambs, while the cougar pounces on any individual.

Three species of goat antelope occur in North America: the Rocky Mountain goat, which ranges from Alaska to northern USA; Dall's sheep of Alaska and northern Canada; and the bighorn, which lives throughout the mountain chain. The Rocky Mountain goat is an exclusive mountain-dweller, perfectly adapted for life high in the alpine zone (see box on page 341). Dall's sheep and the bighorn are closely-related species that spend most of their time at lower elevations, although the bighorn will roam above 12,000 ft in summer.

Clash of the bighorns

Bighorn feed mainly on grasses which they find in the meadows above the treeline. In the spring, the mountains echo with the sound of rival males competing for dominance by clashing their massive, curled horns. The powerful headbutts are a form of ritual fighting designed to exhaust rather than injure opponents. Strong reinforcement in the bones of the skull absorbs the impact of the

Double-Coated Mountaineer

The Rocky Mountain goat inhabits remote mountain peaks from north-western USA to south-eastern Alaska, where it occupies meadows and crags up to the fringes of glaciers. The hardy animal spends much of its life in a landscape that is blanketed with snow, its creamy-white coat providing excellent camouflage. It lives on the scant, coarse vegetation that grows on the upper slopes, and shuns the lower reaches.

The Rocky Mountain goat is protected from the cold by its thick, double coat. The coat has a woolly underfur and long, stiff hair on the back, neck and thighs. An expert climber, it is able to negotiate the most treacherous slopes, leaping the length of its body onto a higher perch or jumping several feet down onto an icy ledge. The Mountain goat's sure-footedness results from the structure of its hoof. It has a flexible pad on the sole that provides tread, and a rim around the hoof that catches onto irregularities in the rock. When walking on snow, the hoof splits into two halves and splays out, spreading the goat's weight. But when walking on narrow ledges, the hoof remains closed. Secure on its four feet, the Rocky Mountain goat is safe from most predators. However, if a congar makes an attack, the goat's sharp, stabbing horns serve as fearsome weapons.

RIGHT The Rocky Mountain goat has flexible hooves that are rimmed with hard edges, enabling the goat to grip the narrow ledges of precipitous mountain slopes. It can leap directly upwards to grasp footholds 3 feet above, and will jump several feet down a steep rockface onto a ledge. Here, a group of Rocky Mountain goats climb a sheer crag in Glacier National Park to lick mineral salts from the stone.

blows, even though the rivals put their full weight behind each charge. Dall's sheep have more slender horns and the males are more lightly built than the bighorns. They also differ in that two of the three subspecies have white coats, providing better camouflage in the snowbound terrain of the far north.

Beauty in the clouds

Below the southern border of the USA, the mountain chain curves through Mexico as the Sierra Madre Occidental, the Sierra Madre Oriental and, further south, the Sierra Madre del Sur. The slopes of the Mexican mountains are mainly hot and dry, with scrubby vegetation. Most peaks are below 12,000 ft but in a few places giant volcanoes reach far greater heights. The volcano Popocatepetl, which is located just outside Mexico City, measures 17,800 ft. The same pattern of moderately high mountains, combined with taller volcanic peaks, continues through Guatermala, Honduras, Nicaragua, Costa Rica and Panama. As the climate becomes more humid the vegetation on the mountains changes in character. Dense mountain forests cloak the lower slopes with leafy tropical plants. Higher up the mountainside, where fog and mist collect on the slopes sheltered from the wind, cloud forest appears.

LEFT With feathers that shimmer when they catch the light, the resplendent quetzal is a spectacular inhabitant of the Central American cloud forests. Swirling mists usually cloak the forests, and in the overcast conditions the bird is surprisingly elusive. It spends its time high in the tree canopy, keeping its bright breast feathers turned away from intruders. In the murky light, the plumage on its head and back closely match the surrounding greenery.

Stunning colors

The cool, humid cloud forests of Central America comprise some of the richest mountain habitats in the world. They harbor a profusion of brightly marked frogs, parrots, monkeys and other animals reminiscent of the hot rain forests found at much lower altitudes. One of the most spectacular residents of the Central American cloud forests is the quetzal. An inhabitant of forests above 5,000 ft, the quetzal spends most of the time in the tree canopy, where it feeds on fruit, insects and the occasional frog. The male has resplendent green upper-parts with a 3 ft long train of green tail feathers and deep red underparts. The feathers are so beautiful that the species has suffered widespread hunting and trapping by local people. In the bird's natural habitat, the colors are much less conspicuous. The green plumage shimmers in the sunlight, but in the overcast conditions of the cloud forest, it becomes remarkably dull. When intruders approach, the male turns his back towards them, and his green feathers provide fine camouflage against the mass of leaves in the canopy.

Continuing south

As the mountain chain flows into South America, the rich combination of mountain forest and cloud forest habitats continue. In the northern Andes, luxuriant forests run up both flanks of the range, reaching as high as 10,000 ft above sea-level. But just south of the Equator they only appear on the eastern side of Amazonia – to the west lies the barren, almost lifeless Atacama Desert.

In many places, the forests cover such steep slopes that it is hard to believe trees can take root. On the eastern flanks of the Andes in southern Peru, the Urubamba river cuts a deep, winding gorge with precipitous slopes. At one section, the river lies in a curving trough rimmed by immense walls, with a towering pedestal inside the river's bend. In spite of the steep terrain, dense tropical vegetation cloaks the slopes, making the valley sides almost impenetrable. The impressive Inca ruins of Macchu Picchu stood forgotten on a ridge alongside the pedestal for over 300 years, until their rediscovery by an explorer early this century.

Birds of the Andes

The dense forests of the northern and central Andes are home to a variety of animals that shun the rain forests below. Hummingbirds are especially numerous, with a variety of species occupying different altitudes. They include the sword-billed hummingbird, the marvellous spatuletail and the black-tailed trainbearer. The forested slopes are also the home of the Andean cock-of-the-rock, in which the male bears brilliant red plumage and a semi-circular crest. The species often inhabits steep, rocky slopes amid a tangle of vegetation, and the female builds her nest out of mud on the rock faces.

The mountain tapir and the spectacled bear are two large mammals that inhabit the high forests. The mountain tapir divides its time between the upper limits of the forest and the open landscape

above, sometimes foraging at night to heights of 14,000 ft. Its thick, woolly coat provides extra protection from the cold. The spectacled bear also has thick fur, and takes its name from the pale hairs that mark its face. Like the mountain tapir, the spectacled bear occurs in both forest and open ground. It usually feeds on plant matter, especially palm nuts and young palm leaves, but also attacks guanacos.

Snowcaps and basins

Above the forested slopes in the high Andes, the landscape consists of steep peaks and broad, flat basins. Through the length of the Andes range there are hundreds of peaks more than 16,000 ft in height, many of which are snow-capped volcanoes. A cluster of volcanoes occurs near the Equator in Ecuador, including Chimborazo (20,000 ft) and Cotopaxi (18,000 ft. Further south, in Peru and Bolivia, stand more giants, including Huascaran (22,200 ft) and Sajama (21,200 ft). The highest of all occur in Argentina and Chile, with Aconcagua (22,834 ft) and Ojos del Salado (21,000 ft). For most of its length the mountain range is narrow – about 100 mi across from east to west. But in the central Andes of southern Peru, Bolivia, northernmost Chile and northern Argentina, the range broadens out to a width of more than 300 mi. Here, the flatter land that lies between the two snow-capped flanks of the range forms a broad plateau called the Altiplano.

The Altiplano plateau

A cold, windswept terrain, 10,000–14,000 ft above sea-level, the Altiplano is a testing environment for wildlife. The land re-

LEFT The jagged peaks of Cerro Torre (left) and Cerro FitzRoy (right, 12,000 ft) accentuate the bleak landscape of the southern Andes. At the opposite end of the New World mountain chain from Mount McKinley, they too lie at cold latitudes, where ice and tundra vegetation spread down towards the lowest slopes.
ABOVE The bleak mountains of the southern Andes are the stronghold of the Andean condor. The largest of the vultures, it has the greatest wing area of all birds of prey. It roosts on the rugged, windswept crags, and soars majestically in the gales, sometimes floating for hours without a single flap of its powerful wings.

ceives little rain, and many areas are classed as desert. Bunch grasses and sedges predominate, with boggy patches occurring around ponds and lakes. Most of the lakes are small, but Lake Titicaca forms a vast, shimmering body of water, about 100 mi in length. In spite of its harsh nature, the Altiplano has long been the focus of settlement, and many cities stand on the flat ground.

Colonies among the rocks

A variety of rodents inhabit the Altiplano grasslands and the mountainsides that rise above them. They include the chinchilla and the mountain viscacha, whose dense fur allows them to range close to the snowline. Both species dwell in colonies among the rocks and forage over the ground for grass, mosses and lichen. Tucotucos dig extensive burrows in the soil, carefully surveying the surroundings to ensure that no danger lurks before they scurry out of their entrance holes.

For reptiles, the Altiplano is a harsh, foreboding place where the temperature is too cold for most species to remain active. But the lizard *Liolaemus multiformis* survives at altitudes of nearly 18,000 ft because it takes great pains to catch the sunshine. At night, it shelters in burrows or beneath rocks, where the temperature does not fall below freezing. In the morning, when the sun climbs into the sky, it emerges to bask, tilting its body so that the sunlight falls as squarely as possible on its skin. It can also change its color, keeping itself dark when basking to absorb more heat. Even when the air remains cool, the lizard can warm its body to 88 F – a sufficient temperature to keep its muscles active during the search for food.

Stronghold of the condor

In the southern part of the Andes, where the mountain chain moves out of the tropics once again, the Altiplano narrows and the Andes simply become a range of jagged peaks. South of the greatest mountain, Aconcagua, the peaks become steadily lower, but vegetation recedes with them as they continue towards the poles. At the southern tip, in Tierra del Fuego, the snowline lies at only 2,300 ft above sea-level. Glaciers move

ABOVE The mountain gorilla inhabits the forested lower slopes of the six extinct volcanoes in the Virunga range of Africa. It has longer, silkier fur than its lowland relatives, enabling it to combat the cooler night-time temperatures. Mountain gorillas eat leaves, shoots and stems that they tear from vegetation with their hands. They feed mainly on the ground, usually in clearings where there is plenty of vegetation within reach.

directly into the sea and the slopes are cold and barren.

In the desolate landscape of southern Chile, the Andean condor is at its most majestic. An enormous vulture, it has the largest wing area of any living bird, with about two square yards of dark flight feathers. Outstretched, the wings enable the condor to soar for hours on end without flapping, and it can maintain its course against howling gales that rip through the mountain heights. Like all vultures, it is a scavenger that scans the ground from on high, on the look-out for carrion. In many places their numbers are low, but in the mountains of Chile, where the peaks of the New World head southwards to the sea, Andean condors continue to thrive, relatively free from disturbance.

Heights of Africa

Africa is a rugged plateau about 1,700 ft above sea-level that contains few true mountains. It is composed of lines of hills and escarpments, and broad valleys separating flat plains. Its truly mountainous regions form only isolated areas on the map.

The continent's highest points lie in East Africa, where broad stretches of savannah separate ranges and volcanic peaks. To the west, on the fringes of the lowland rain forest, stand the Virunga and Ruwenzori ranges. Running along the Zaire border from Rwanda to Uganda, they reach their maximum height at Mount Stanley (16,760 ft). Five hundred miles to the east stand the Mau Escarpment and the Aberdares of Kenya, the latter almost reaching 15,000 ft. In Kenya, Tanzania and Uganda, isolated volcanoes rise dramatically from the plains. They include Mount Elgon (14,000 ft) and Africa's two greatest summits – Mount Kenya, or Kirinyaga (17,000 ft), and Kilimanjaro (19,300 ft).

Located close to the Equator, the East African mountains display a diversity of habitats. Hot, dry savannah at the foot of Kilimanjaro changes to cool, humid forest on the middle slopes and snow-fields and glaciers occur at the top of the mountain. The changing climate and the succession of differing habitats upslope, strongly influence the pattern of wildlife on the mountains.

Forest havens

Dense mountain forests cover the lower slopes of the East African mountains, and narrow columns of bamboo predominate at about 10,000 ft. The forests provide a haven for many lowland jungle creatures, including colobus monkeys and the bongo (a type of antelope). The thin white stripes in the bongo's fur serve to break up its outline against the twigs and foliage in the dimness of the forest floor. Leopards prowl the mountain forest, putting their tree-climbing skills to good use, and bushbuck forage among the trees, ever watchful for the threat of attack from leopards hiding in the branches.

Some animals are specialized for life on the forested slopes, rarely appearing in lowland trees. The Tacazze sunbird and Jackson's francolin are two birds that live only in mountain forests above 7,000 ft. The mountain gorilla is perhaps the most celebrated resident of the slopes. A subspecies of the gorilla, it is restricted to mountain forests of the Virunga range, where a few hundred individuals have survived the relentless hunting that endangers the ape. Gorillas spend most of their time on the forest floor, pulling leaves and fruit from branches close to ground level. They often occur in patches where the vegetation is low or where there is a break in the forest canopy. In such locations, plenty of light reaches ground level, encouraging the development of dense, leafy undergrowth.

Giants in the mist

The mountain gorilla frequently ranges above the treeline to feed on the grasses and shrubs on the high, misty moorlands that stretch towards the summit. Here, in the realm of giant tree heaths, tree groundsels and vastly outsized lobelias, other animals that normally appear lower down occasionally come to forage. Elephants climb to 14,000 ft on Mount Kenya, where they feed among the heaths and chew the leaves of the groundsels. Eland range to similar heights, where they are thought to lick salt from the bare rocks below the summit.

Rock hyraxes are particularly at home in the alpine moorland, where they nibble the grass tussocks and gnaw the bases of lobelias. They take shelter among the rocks, where the flexible pads on the soles of their feet allow them to clamber over rocks and stones with ease. Ranging as high as 15,000 ft, they face extreme cold at night when the temperature dips below freezing. To combat the chill, they have longer hair than their lowland counterparts, but even with such protection they are in danger of losing heat. At night, small groups of the animals huddle together in crevices among the rocks to keep one another sufficiently warm. When the morning sunshine returns, they lie on exposed surfaces, basking in the warm rays.

ABOVE Above about 10,000 ft, giant tree heaths grow on the misty, high-altitude moors of several East African peaks. Mosses and lichens festoon their slender trunks and branches. The giant tree heaths are relatives of the heather that grows in many parts of Europe and North America. While northern plants grow to 2 ft, their African mountain cousins often measure ten times this height.

ABOVE The scarlet-tufted malachite sunbird lives high in the alpine zone of East African mountains where it feeds on the giant lobelia. The alpine plant produces an enormous flowering column up to 20 ft high, with a mass of pointed leaf bracts. The sunbird perches on the bracts to sip the nectar from the flowers in between. In return, the sunbird acts as pollinator for the giant lobelia by ferrying pollen from one plant to another.

Exclusive residents

A few species inhabit the high moors of the East African peaks. Two burrowing mammals – the Mount Kenya mole rat and the Mount Kenya mole shrew – live only on Mount Kenya, while the montane francolin is scattered across the high peaks of East Africa and Ethiopia. The scarlet- tufted malachite sunbird – a brilliant metallic-green species with very long tail feathers – lives in close association with the giant lobelias. By perching on the leaf bracts, and probing between them with its slender bill, the sunbird can reach the mass of flowers within the ten foot high plant. As it sips the nectar, it transfers the pollen that dusts its bill.

High hopes

The difficulties of life on the mountains affect people just as much as wildlife, and in many cases the harsh conditions have saved them from the destruction that so many lowland habitats have faces. Because of the cold climate, the thin air and the steep terrain, humans have been discouraged from settling on the high slopes. Mountains therefore represent some of the our last great wilderness areas, providing a precious refuge for wildlife. In Europe, mountains are the final strongholds for wolves that have been driven out of the lowlands. Wild populations still exist in the mountains of Spain and Italy, in eastern Europe, and in the Caucasus and Urals of the USSR.

In many parts of the world, mountain environments have gained national park status in recognition of the natural riches they still contain and the need to preserve them from future threats. Mount Kenya National Park has become an island of natural diversity within a sea of cultivation. Farms and plantations run right up to the mountain, but at the border of the Park the natural forest takes over.

Vulnerable giants

Not all mountains have escaped the effects of human settlement.

People have inhabited the flat Andean Altiplano for thousands of years, steadily depleting the vegetation cover and converting the land to pasture. The Alps have experienced a long history of settlement in the deep, flat-bottomed and accessible valleys that Ice Age glaciation left behind. Trees have disappeared from vast tracts of mountainside and villages cling to the slopes. Chamois survive in many parts of the Alps, but few other large animals have been able to adapt to the changes to their environment.

The Ethiopian Highlands demonstrate how vulnerable mountains can be to the impact of human settlerment. Once well vegetated, great sections of the range are now bare and desolate. Clearance of the vegetation for firewood and overgrazing by domestic livestock stripped the land of its soil cover. Thin and easily washed away on the steep slopes, the soil disappeared, leaving behind a dry wasteland. Soil erosion is one of the greatest problems facing the poor countries of the world, and in few places does it proceed more rapidly than in tropical mountains. Despite the adverse conditions at high altitude, the pressure of population and dwindling resources in the lowlands is pushing people further and further upslope.

RIGHT In many mountainous regions of the world, land clearance and farming have drastically altered the mountain environment. Intricate terracing is a common sight in the foothills of the Himalayas and on mountain slopes over much of east and Southeast Asia. By tracing the contours of the mountains, level terraces are created for intensive rice cultivation.

CAVES

Dark, damp and forbidding, caves seem to be one of the least welcoming of habitats, yet they contain a surprisingly varied community of creatures, some of which use the caves for temporary refuge, others as permanent homes

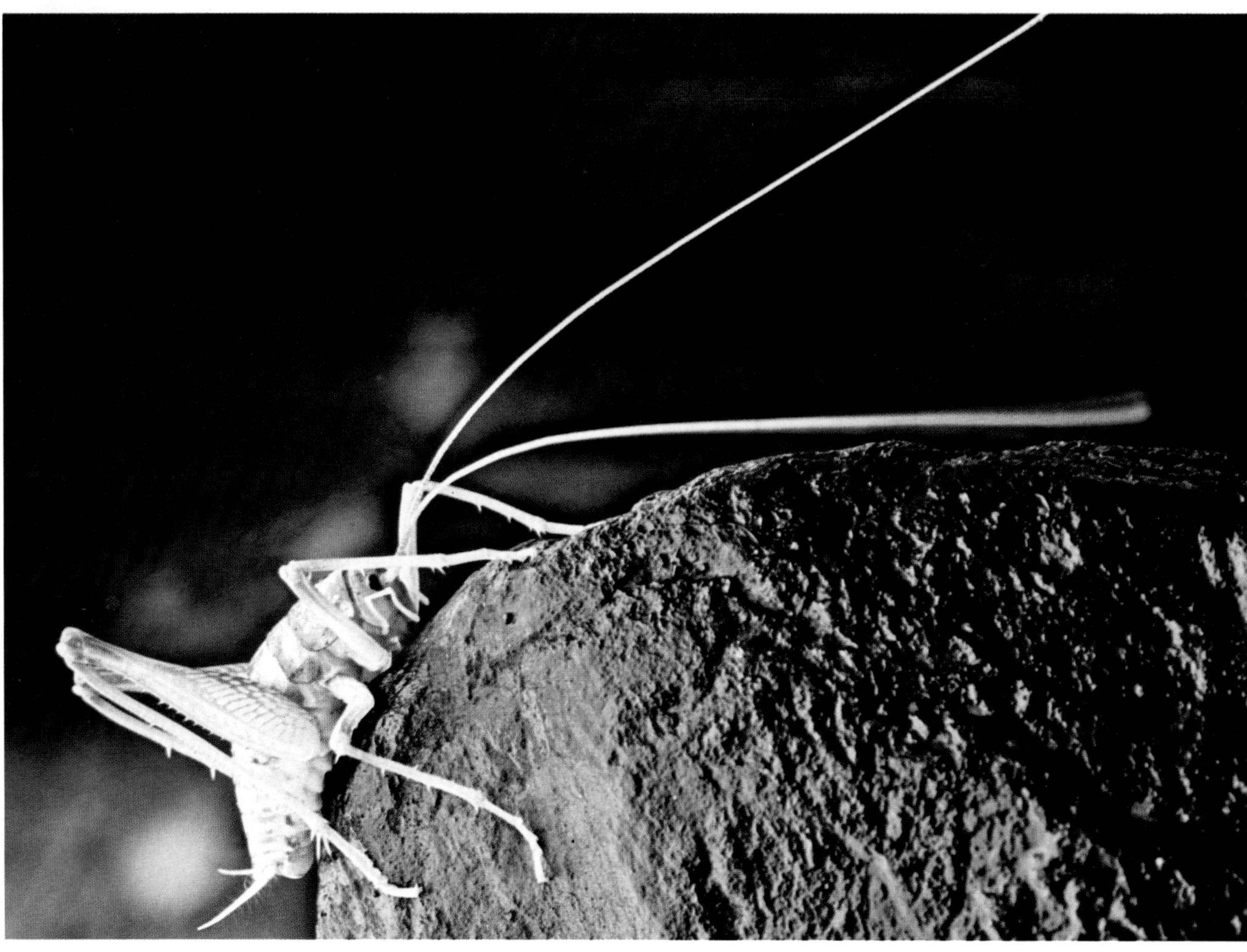

ABOVE The darkness of caves has prompted some extraordinary adaptations among animals that make the habitat their home. Cave crickets are common scavengers that search the dark recesses for carrion, such as the dead bodies of bats. The crickets' elongated antennae may reach more than four times their body length. By contantly waving them, the insects feel their way through the passages, scan the rock for food and gain an early warning of approaching predators.
PAGES 350–351 The total absence of sunlight in a subterranean cave prevents the growth of the green plants that sustain most food chains. Yet, even caves have their own distinct wildlife communities. All cave animals are dependent on scraps of food that are washed underground by rainwater or brought into the caves by animals that venture outside.

No sunlight penetrates into the depths of a cave. Beyond the entrance tunnel, the subterranean realm lies in perpetual darkness. But in the artificial light from a caver's lamp, the cave reveals its secrets. Twisted passages and gaping shafts appear, widening into hollow caverns with rushing streams and pools of water that are as clear as glass. Strange, ornate formations – deposits of crystalline rock – dangle from the roof or rise from the floor, their smooth, damp surfaces glistening in the light. The cave is a world apart from the wildlife habitats outside, but, even here, animals thrive, adapted to a dank, dark environment unlike any other on earth.

Hollows and tubes

Caves develop in a variety of landscapes and are formed in a number of ways: the constant battering of the sea erodes short tunnels in coastal cliffs; strong winds armed with sand grains carve hollows in the sides of hills or banks; and streams and warm winds melt passages through the ice of glaciers. They are also formed during volcanic eruptions, when rushing torrents develop deep inside streams of lava. But the greatest and most extensive cave systems are those that form in limestone as rain-water percolates through its cracks and crevices.

Limestone is a common rock that occurs in many parts of the

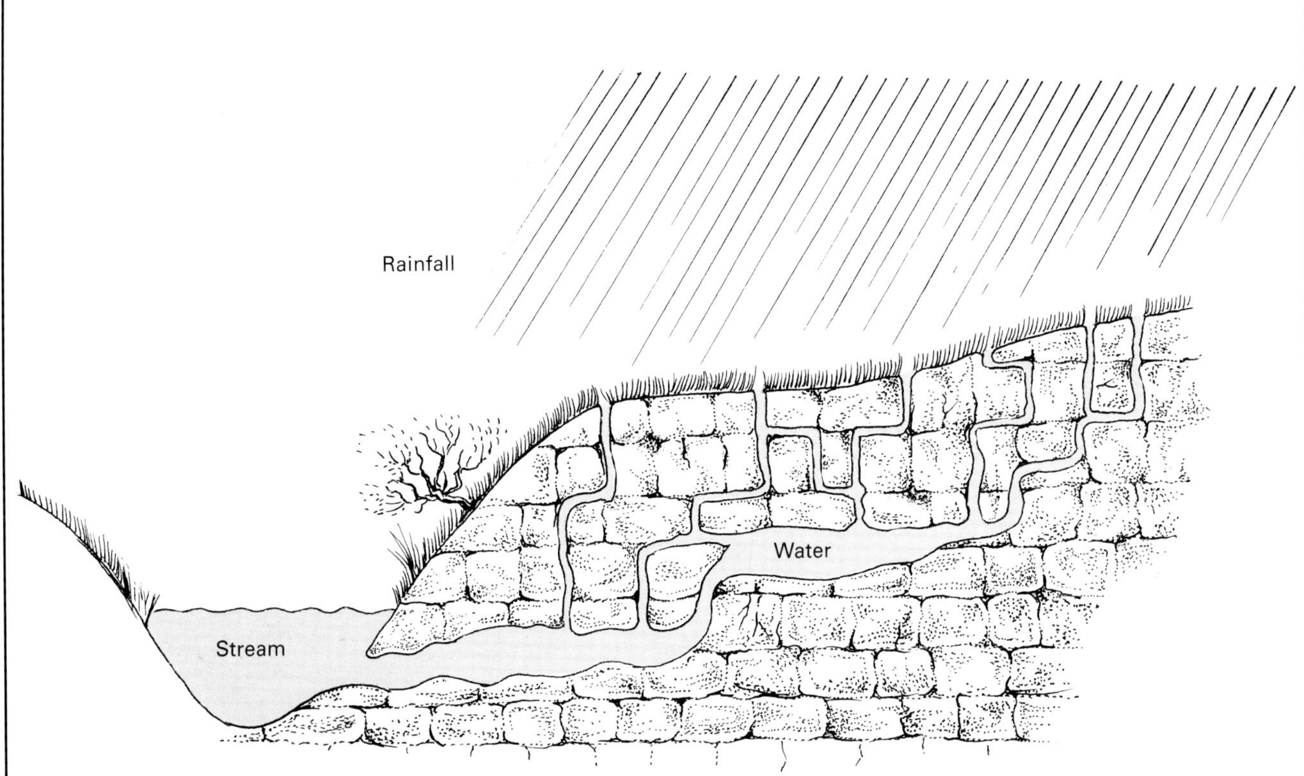

ABOVE Most caves from when rainwater steadily percolates through limestone hills. Limestone contains a network of joints that form weak points in the rock. As rain falls, it seeps into the joints and filters downwards, dissolving particles of the limestone. The joints widen to form fissures and eventually a system of shafts and tunnels. As water continues to flow along the passages, the main channels are enlarged, forming broad caves. The underground streams emerge as springs or flow directly into rivers.

world. It began to form millions of years ago, when the hard fragments of dead shellfish, corals and other marine creatures accumulated on the beds of ancient tropical seas. The fragments – composed mostly of the chemical calcium carbonate – formed thick layers. As more and more material settled above, weighing down on the fragments below, each layer steadily compacted and hardened until it turned into solid rock. Later, earth movements lifted the limestone up from the ocean bed to form dry land. Wherever limestone appears near the surface today, caves are likely to occur.

Extensive limestone regions exist in Eurasia, Africa, Australasia and the Americas. In Europe, broad belts of limestone occur in the Alps and in Yugoslavia, with scattered patches across the remainder of the continent, including Spain, southern France, the British Isles and European USSR. The great cave of Pierre St. Martin in the Pyrenees is one of the world's deepest caves – its lowest point is 4,370 ft below its highest entrance.

Limestone regions reappear in many parts of Eurasia, including a vast region of southern China. In Borneo, the spectacular Mulu caves contain the largest single cavern in the world. Called the Sarawak chamber, it measures 700 yds long, 300 yds wide and at least 200 ft high. New Guinea, southern Australia, the Horn of Africa and Andean South America all have extensive cave systems, but the greatest of all lies in the USA. The Flint-Mammoth system of Kentucky contains over 220 miles of passageways.

Widening fissures

Limestone caves are created when rainwater reaches the surface of the rock. Tensions in the earth's crust create thin cracks or joints in the limestone that are at right angles to one another, allowing rainwater to seep steadily into the joints. Having passed through the soil, the water contains a store of carbon dioxide released by decomposing plants. High levels of carbon dioxide greatly increase the rate at which water dissolves the calcium carbonate in limestone. As the water seeps through, it dissolves the rock around the joints, creating narrow fissures.

Over thousands of years, the fissures steadily widen, and the flowing water creates a system of passages that run down through the rock, twisting and turning at right angles as they follow the pattern of the original joints. In most cases, one route through the rock predominates, taking most of the water flow and becoming greatly enlarged in the process.

Cave passages that are completely full of moving water are usually tubular in shape. The water dissolves the rock equally in all directions, leaving smooth, rounded walls. But in many caves the water level eventually drops, emptying the passages and allowing air to pass in. Water still percolates downwards, but now it gathers in pools and flows along the cave bottom in streams and waterfalls. In such cases, the water enlarges only the cave bed, eroding a gulley in the floor. Some caves drain almost entirely so that only the sound of moisture dripping from the roof reveals that water is present in the rock.

Crystalline decorations

When water seeps to the roof of a cave through narrow fissures, it contains high levels of carbon dioxide and dissolved limestone. As it comes into contact with the air that fills the passage, much of the carbon dioxide bubbles away into the subterranean atmosphere. As the carbon dioxide departs, the water steadily loses its power to hold calcium carbonate in solution. Some of the dissolved material becomes solid once again and forms hard crystals of calcite.

Over time, impressive formations of calcite build up, stained a variety of colors by other minerals present in the rock. Where water drips slowly from the roof, downward-pointing stalactites develop in the shape of needles or cones. Underneath, thicker, conical mounds called stalagmites grow upwards from the floor where the falling drops deposit more of their load. Stalagmites may grow to an enormous size in caverns – those in the Aven Armand cave of France measure 100 ft in height. After perhaps 100,000 years, stalactites and stalagmites may join to form tall columns. Where water seeps across a slanted floor, layers of "flowstone" may develop, and where it runs along a slanted roof, it commonly forms delicate, curtain-like projections known as "cave drapes." When light shines on the glistening calcite formations, remarkable displays of shape and color are revealed.

Constant conditions

The strange subterranean world provides an extraordinary habitat for wildlife. Far from the sunshine, wind, rain and snow, deep caves have stable climates barely affected by the changing seasons on the surface. Over 100 yd from the cave entrance, the air temperature fluctuates no more than a few degrees throughout the year. Deep inside the Flint-Mammoth cave system, the temperature is always between 57 F and 58 F. Most caves are continually humid, and with moisture saturating the air, the walls are damp to the touch. Though traces of reflected and filtered sunlight may reach for more than 100 yd along straight entrance tunnels, deep within caves the darkness is complete. Nocturnal surface-dwelling animals have large

ABOVE Each time a drop of water collects on the roof of a cave or falls to the floor, it deposits a tiny piece of dissolved limestone as a crystal of calcite. Water that drips slowly from the roof eventually forms "stalactites" or hanging spikes. Underneath, thicker conical mounds called "stalagmites" grow upwards from the floor where the falling drops deposit more of their load. They are stained a variety of colors by other minerals present in the rock. Over many thousands of years impressive crystalline formations build up on the roofs, walls and floors of caves, including large columns where stalactites and stalagmites have joined together.

LEFT Where water seeps across a slanted floor, layers of "flowstone" may develop, and where it flows along the slanted roof of a cave, delicate, curtain-like projections known as "cave drapes" are formed. When light shines on these calcite formations, they glisten, revealing remarkable displays of shape and color.

LEFT A variety of animals divide their time between caves and surface habitats. Bears may enter caves to find shelter during winter, and rock wallabies seek their shade during the heat of the day in Australia. Bats and swiftlets roost and nest in caves, but catch their food in the air outside, and some crickets make regular foraging trips to the surface. Some animals, such as the New Zealand glow-worm, spend their early stages inside caves, but live their adult lives in the forests.

eyes to gather the dim light from the moon and the stars, but for deep cave-dwellers there is no source of light at all.

Underground, food is scarce, and with no light available for photosynthesis, green plants never grow. Caves contain some bacteria that produce organic matter from silt and from the limestone itself – without the need for light – but, for the most part, the cave community relies on debris reaching the cave from outside. Streams and percolating water bring dead vegetation and other detritus into the cave, particularly during floods. The debris provides nutrients for decomposers such as fungi, worms and millipedes. Small animals drift inside, and any cave creatures that regularly forage outside for food provide nutrition in their faeces. The cave animals that process dead plant and animal tissue themselves provide food for predators.

Cave communities

Though caves generally contain far fewer species than their immediate surroundings on the surface, a diversity of creatures inhabit the dark passage, and some never stray outside. The majority are small invertebrates – mainly arthropods – but larger

WINGS IN THE CAVERNS

Throughout the temperate and tropical regions of the world, the subterranean habitats of caves provide a haven for bats. Horseshoe bats, vampire bats, tomb bats, free-tailed bats and flying foxes all shelter in caves, and some gather in spectacular numbers. Tamana Cave in Trinidad accommodates 12 species of bats, each present in their thousands. Every evening, 300,000 free-tailed bats leave Deer Cave in the Mulu caves of Borneo, creating a roar of beating wings audible two miles away.

Bats crowd into caves in such numbers because the walls and roofs offer secure roosting sites. They gain shelter from the wind and the rain and are out of reach of most predators. During the breeding season, cave walls provide sites for maternity wards, where females tend their young in safety. In temperate climates, they provide a secure refuge for hibernation. Hanging from their hindfeet in clusters of up to 3,000 bats per square yard or wedged into crevices in the rock face, the animals enter a state of torpor that lasts throughout the harsh winter season.

Seeing in the dark

The absence of light in the caves presents few problems for bats. With their sensitive echolocation skills, they can navigate with ease past the walls and hanging stalactites in total darkness. By emitting high-pitched sounds and listening for the echoes, they gain an accurate mental image of their surroundings as they race through the tunnels to and from their roosts.

BELOW A long-eared bat, with its newly-born young attached to its underside, flies through the darkness of a cave to a new roosting site. Caves provide a secure refuge for colonies of bats throughout the world – havens where the animals rest, hibernate and rear their young.

LEFT Navigation through the narrow passages of dark caves is difficult for cavers even with artificial lights. But cave-roosting bats and birds have to fly through the tunnels in total darkness. To guide their way, they emit pulses of sound and listen for the reflection of the sound waves from the cave walls, a method known as echolocation.

inhabitants include fishes, salamanders, birds and bats.

Bats – the animals most popularly associated with caves – occur in caves throughout the world. They spend the day roosting on the cave roofs and walls, secure from danger. At dusk, they fly outside in search of insects or fruit (see box on page 357). Since they eat more than half their body weight of food in a single night, bats produce great quantities of faeces or "guano." It falls to the cave floor, providing rich feeding grounds for other creatures. Rhaphidophorid crickets perform a similar function in many North American caves. They regularly forage outside at night, and when they return, they produce copious dung – a layer half an inch thick carpets the cave floor, providing food for beetles.

Creatures that find all the sustenance they need underground belong to simple food webs, each associated with a different part of the cave. Flatworms, isopods and tiny white shrimps provide food in streams for water beetles, crayfish and the specialized cave-fishes of North America. Springtails, beetles, cockroaches and crickets inhabit the drier portions of the caves, providing food for spiders and salamanders.

Tropical hordes

The caves of tropical Borneo contain vast colonies of bats – often containing over 500,000 individuals – and swarms of swiftlets that hunt over the forests for insects by day and return to the caves at night. Swiftlets breed on the cave walls and roofs – sometimes hundreds of yards from the entrances – mixing moss with saliva to mold cup-shaped nests against the rock. The bats and the birds produce huge masses of guano that form stinking dunes on the cave floor. The guano is usually alive with golden cockroaches, and beneath the swiftlet nests, moths' larvae forage through the litter. Each larva has a protective casing of old insect cuticles surrounding its body. Dying bats or birds that fall to the floor quickly attract the attention of cave crickets and freshwater crabs that scavenge on the bodies. The chief hunters of the cave floor include large huntsman spiders and giant, poisonous centipedes that measure up to 9 in in length.

Senses in the dark

The lack of light, the abundance of moisture and the scarcity of food in caves have had a marked impact on

the sppearance and behavior of cave wildlife. Many cave inhabitants lack skin pigmentation and many species of spider, millipede, shrimp, fish and salamander all share a white, ghostly color. With no light available, they do not need to camouflage themselves or display breeding colors. The manufacture of pigments in their skin would be a waste of precious energy. Similarly, cave creatures have no use for sight. Many are blind, and some, like the olm and the Texas blind salamander, have a sheath of pale skin covering the remnants of their eyes.

In the absence of eyesight, cave creatures have evolved exceptional skills in other senses, especially in touch. The texas blind salamander has sensitive nerve endings on its head that detect vibrations in the water. The southern cavefish – a blind and almost colorless cave species – relies entirely on the row of sensory organs that make up its lateral line to detect tiny crustaceans in the water. It has

TOP The oilbird roosts and nests on the walls of South American caves, sometimes as far as a half mile from the cave entrance. Emitting rapid tongue clicks, they navigate their way through the tunnels by echolocation. In caves with large colonies, the noises from thousands of birds racing through the passages can be deafening.

ABOVE Cave centipedes are fearsome predators on the floors of tropical caves. Reaching 9 in in length, they have long legs which enable them to scuttle rapidly across the rock surface. They often lie in wait for prey such as cave crickets to draw near. When they brush past, the centipedes seize their victims and kill them with their poison claws.

ABOVE The New Zealand glow-worm – the larva of a species of fungus gnat – produces lines of silken thread, each armed with droplets of sticky fluid. The larva glows with the light produced by luminescent bacteria in its gut. In the darkness of ravines and caves, the light illuminates the dangling threads, attracting flying insects. Some become ensnared on the sticky lines, and the larva draws its victims up for digestion.

four times as many sensors as its close relative, the springfish, which mainly lives on the surface.

Many cave-dwellers have developed long antennae, enabling them to detect food and predators at an early stage. Some cave crickets have enormous antennae over 15 in long and many times their body length. They constantly wave the slender probes until they come into contact with food, such as the body of a bat on the cave floor.

Many of the larger inhabitants use sound to navigate through the cramped cave passages. Oilbirds nest deep inside South American caves and emit streams of sharp tongue clicks as they fly through the darkness. The pattern of echoes from the cave walls, from stalactites and from other birds, enables them to build up an acoustic image of their surroundings. Swiftlets and bats use a similar system. Scientific evidence suggests that bats memorize the image of their home so that they do not always need to listen out for echoes – they often collide with artificial objects placed in caves.

Though food sources are sometimes rich and concentrated in caves, such as in the piles of guano beneath bat roosts, most cave creatures have difficulty in finding food. As a result they have become less specialized in their diets than their surface-dwelling counterparts and will eat almost any type of food that becomes available. Beetles and mites often double as scavengers and predators – swarms of mites will regularly attack breeding colonies of springtails. Nothing goes to waste in the caves, and creatures are extremely resourceful in their quest for food. Giant earwigs inhabit the roofs of tropical caves, where they pick dead skin from roosting bats. Blind rhadinid beetles search for the eggs of crickets in silt, piercing them with their sharp mandibles. Cave racer snakes, measuring up to six feet in length, hunt for dying bats on the floor. They will climb the walls, if there are sufficient irregularities in the rock, to pick off roosting bats and swiftlets.

Fragile habitat

As more and more people explore caves, our knowledge of the habitat continues to grow. Many cave systems have yet to be mapped, and new discoveries regularly occur. In 1978, the crustacean *Speleonectes lucayensis* was discovered. It lives in an underwater cavern in the Bahamas, and is so different from any other crustacean that zoologists placed it in a class of its own. But as caves yield their secrets, so their formations and their wildlife become more vulnerable. Broken stalactites show that souvenir-hunters have been at large. Many bats cannot tolerate disturbance and will abandon cave colonies all too easily, removing the vital source of guano that other creatures rely on. Cave habitats contain fragile, almost self-contained communities and with thoughtful conservation, we still have a chance to preserve them.

The little egret (*Egretta garzetta*) lives in warm regions of the Old World wherever shallow freshwater occurs. It probes for frogs and small fishes among the submerged aquatic plants of pools and marshes, jabbing at its prey with its dagger-like black bill.

RIVERS AND LAKES

From cold, rushing headwaters through broad meanderings to the sea, the river shows many faces, with animals and plants attuned to its relentless flow. Lakes are placid, shimmering habitats that may be glacially clear or rich in vegetation, insects, rodents and waterfowl

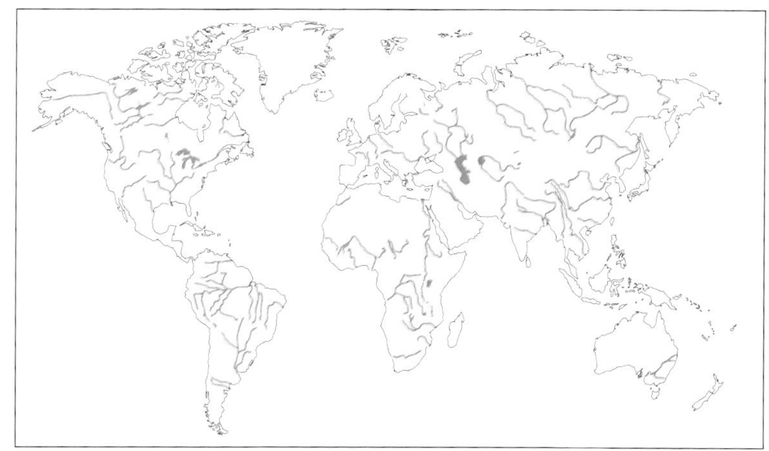

ABOVE Salmon, such as these silver salmon in Alaska, swim upstream to spawn in the shallow headwaters of rivers.
PAGES 362–363 Mountain lakes are freshwater habitats that appear to be still and tranquil, but beneath their calm surfaces, they often teem with life – from microscopic animal plankton to large predatory fishes.

Freshwater throughout the world supports a myriad of life forms in tiny mountain streams, wide, muddy rivers and swamps. The animals of these habitats range from the huge beluga sturgeon to the tiny three-spined stickleback; from the brilliantly colored kingfisher to the dull brown, secretive water rail; and from countless mosquitoes to the jaguar. All are specially adapted to spend their lives in or near water.

There are countless streams, rivers, ponds and lakes in the world, although in total volume they account for only a tiny fraction of the world's water – less than half of one per cent of the total. (Oceans contain about 97 per cent of the water; glaciers and ice caps in polar regions and on mountain tops contain two per cent; and slightly more than 0.5 per cent is held in the soil and in the rocks beneath.)

Plants grow in abundance in rivers and lakes. Some live fully submerged, whereas others float on the surface. Animals may live among the water plants, using them for shelter and for food, or they may bury themselves in the sediment and rocks on the bottom, while a few live on the water surface itself. For some animals, the water provides a permanent home, whereas for others it is only one part of their habitat. Insects, such as dragonflies, spend their early lives in the water, but leave it when they are adults, whereas fishes, such as trout and bream, never leave their freshwater home.

Rivers and lakes worldwide

All the continents have their famous rivers and lakes. Some of the most well-known in Europe include the Rhine, Rhone, Volga, Danube and Thames. The temperate climate, with its regular rainfall throughout the year, provides ideal conditions for the formation of permanent lakes and artificial reservoirs.

Asia contains Lake Baikal (USSR), the Yangtse Kiang (China) and the Mekong River of South-east Asia. Africa has Lake Victoria and the longest river in the world, the Nile. Australia has the Murray and Darling Rivers; North America contains the Great Lakes, and the Mississippi and Missouri Rivers. South America is home to the second longest river in the world, the Amazon.

Salt lakes are typical of the arid parts of the world. All lakes contain some dissolved minerals, but when the concentration of salts in the water exceeds three per cent by weight, the lake is classed as a salt lake. Such lakes include the Dead Sea in Israel, Mono Lake in California and Lake Natron in the Great Rift Valley of Africa.

The water cycle

The freshwater that flows in rivers and fills the lakes is part of a continuous water cycle. It begins as the warmth of the sun turns standing and running water into water vapor. In this process, water evaporates from many sources, including rivers, lakes, seas, the upper layers of soil and the surface of plant leaves. As the vapor rises, it cools and condenses around microscopic particles in the air to form water droplets. Millions of droplets come together to form clouds. The droplets in the clouds then form rain or snow and fall to the ground, completing the cycle.

ABOVE Waterfalls are perhaps the most dramatic features on a river's journey to the sea. Here, the Iguacu River splits up into a series of falls along the border between Brazil and Argentina. Totaling over two miles wide, the Iguacu Falls tumble about 250 ft over an inland cliff, where tropical vegetation grows on precarious overhangs and ledges between the plunging sheets of water. Although waterfalls prevent aquatic animals from traveling upstream, they also have a positive effect on the wildlife of a river. Animal and plant material that is swept over them, for instance, provides food for wildlife that lives downstream. Some birds even take advantage of the protection afforded by the wall of falling water by nesting behind it.

In some cases, the returning rainwater turns to ice and is locked up for many years in the icefields of the polar regions, at the tops of mountains or in glaciers. Some of the rain sinks deep into the ground, joining vast groundwater resources. Later, it may rise to the surface at a spring, or help to fill a river or a lake.

On the hills, tiny trickles of melting snow and rain may merge to form a small stream. As more water joins it, the stream becomes larger and wider. At first it runs between plants and other obstacles, but as it becomes stronger, it starts to cut a channel through the soil towards the rock that lies beneath.

Shaping the land

A river does not simply follow the form of the land; it actively shapes it. Gathering more and more water as other streams join it, it cuts into the land and creates a valley. It dissolves mineral salts from the rocks over which it flows and dislodges larger particles of rock. These scour the banks and the ground, helping to shape the valley of the growing river.

During its journey from the hills to the sea, the river's character gradually changes. At first it is shallow, clear and cool. The bottom is stony, and boulders break the surface. Waterfalls, still pools and sections of racing, turbulent water appear as common features. Tributaries that join the river bring not only more water, but also silt and plant and animal remains. As the river flows off the hills and across low-lying land, it forms wide bends. The river tends to become deeper and muddier where it flows over deep, soft soils that it wears down and carries off.

As it approaches the sea, the river meets the incoming tides, which cause it to slow down. The flow is no longer powerful enough to carry the river's load of sediment, and it deposits its burden at the convergence of the river and the sea. With the sediment, the river builds its own flat land around its mouth. Where the tides and currents are not too strong, the river deposits its sediment out into the sea, forming a delta. In the last stages of the river's journey, it wanders across the delta, splitting and rejoining. Where it meets the sea at the end of its journey, the river often forms mud-flats, swamps, or salt-marshes.

How lakes are formed

Lakes occur in a variety of shapes and sizes, and may originate in a number of ways. The powerful force of glacial erosion, for example, leaves many deep gouges in the land. When these fill with water, lakes are formed. Some lakes have formed in the craters of extinct volcanoes, and others lie where valleys have been dammed by rockfalls. Movements of the earth's crust, where immense blocks of rock have moved vertically and horizontally against each other, have generated basins that now contain lakes. A large number of lakes are man-made.

Freshwater ecology

The animals and plants that live in a river or lake must be able to deal with the water's flow rate, turbulence, temperature and clarity, as well as the nature of the bed and banks. Throughout the world, hill streams, lowland rivers and lakes all have their distinctive fauna and flora. The freshwater environment is a complicated one, and in most cases, the reasons why different species occupy different freshwater habitats are not clearly understood.

Life near the source

In the early stage of a river's journey to the sea, the range of animal and plant species found in and beside it is not very great. The water of young streams tends to be cold, and very fast in places. Heavy downpours can cause a sudden and powerful rush of water. The bed is often rocky, and there is rarely enough sediment for aquatic plants to take root. In addition, a young mountain stream as an ecosystem does not have as layered a structure as, say, a wood or a grassland. For this reason, there are fewer habitats for animals to occupy.

Away from its source, the river picks up organic material from the plants along its banks and from the waste products and dead bodies of its animal occupants. Such debris, known as detritus, forms the basis of life for aquatic animals further downstream. Near the start of the river, such material is in fairly short supply, and only a few animals obtain enough to survive.

Clinging to the rocks

The species that live near the start of the river, including invertebrates such as the larvae of blackflies, mayflies and caddis flies, cope with the fast current by clinging onto rocks and stones. Some of them survive by grazing on the thin smear of algae on the rocks, whereas others trap what little there is in the way of food particles that swirl past them on the current.

Surviving in rushing water

In the young river, invertebrates often congregate on rocks on the river bed where the rushing force of the water is least felt. As the water travels over the stream bed, a thin layer is held stationary against the bottom by the effects of friction. Above this layer, the water rushes past at great speed. The invertebrates take full advantage of the shelter of this thin layer of water in order to avoid being swept away. Many of these creatures are flat, while others have developed suckers or claws to keep them steady.

Wherever the stream bed is composed of sand or gravel, some animals make their home beneath the surface. Mites and other small

ABOVE New Zealand's Rees River wanders through its valley, flanked by the stony material it has carried down from the hills. Like all rivers, it erodes the land over which it passes, carrying down fragments of rock. A river's power to erode is greatest where its flow is strong and swift. Where it flows more slowly, or when its waters are low, a river loses its strength, and deposits its load on the riverbed.

invertebrates live in this dark world, where water filters down to them through the rocks, bringing tiny particles of organic matter for them to eat.

Although the young stream may not support a great variety of animals, the species that do occur can be abundant, and larger animals exploit them for food. Fishes, such as trout, and birds, such as dippers, are typical inhabitants of the upper reaches of streams, and feed on the rich supply of invertebrates that live there.

ABOVE Although some water plants grow rooted in the beds of rivers and lakes, others float freely, their roots hanging down into the water. Certain species of water ferns, such as the *Salvinia natans* seen here, are free-floating plants that can completely cover and clog waterways.

Grazers and shredders

River invertebrates feed in a variety of ways. Water beetles, for instance, graze on aquatic vegetation, while mayfly larvae scrape algae from the surfaces of rocks. Blackfly larvae, on the other hand, collect detritus as it flows past them. Some invertebrates, such as caddis fly larvae and several species of crayfish, feed on the leaves that fall into the water. They are known as shredders, and they eat away the parts of the leaf between the veins. These creatures obtain part of their nutrition from the leaf material; the rest comes from the fungi and other micro-organisms that have colonized the leaf in order to begin the process of decomposition.

The scavenging and herbivorous invertebrates are the potential prey of the river's carnivores. These include larger carnivorous invertebrates, such as the larvae of dragonflies and water beetles, and fishes, such as trout and grayling.

Drifting life

The detritus that the current transports downstream is called "drift." It includes plant fragments, small animals such as worms, tiny snails and freshwater shrimps, and the waste products of aquatic creatures. Much of it is eaten by fishes. Freshwater shrimps release themselves voluntarily from rocks or vegetation into the current to float downstream. In this way, they can spread along the river and exploit new feeding grounds. By drifting at night, small invertebrates are less likely to fall prey to fishes such as trout, which hunt by sight.

The underwater forest

Where a river bed is stony, only a few plants manage to take root and grow. But where finer material accumulates, roots can take hold and a wider range of plants survive, creating an underwater forest of water-weeds.

Plant species that live in water display a variety of adaptations to their habitat. Some grow rooted to the bottom and keep all their leaves below the water surface. Others, also rooted to the bottom, have leaves whose air cells keep them afloat on the surface. Some species have both floating and submerged leaves, while other water plants live a completely free-floating existence, their roots dangling in the water.

At the water's edge, reeds and similar plants put down roots below the water, but their stems and leaves emerge above it. Such "emergent vegetation" is the dominant form of plant life in swamps and marshes and at the edges of rivers and lakes, and provides an important link between the aquatic and terrestrial worlds. For example, warblers nest among emergent vegetation, and dragonfly larvae clamber up their stems so that the adult insects can emerge into the open air.

Certain aquatic plants produce their flowers above the water surface, whereas others flower underwater. The seeds of some water plants float, so that the

current washes them downstream, enabling the species to colonize new sites at a greater distance from the parent than would otherwise be possible. Many water plants grow from fragments of root and stem, as well as from seeds. When these fragile plants are broken up by water birds, animals or passing boats, the bits and pieces of the plant float away with the current, and eventually take root in a suitable site downstream.

Design for living

Unlike plants that live on dry land, aquatic plants do not have to fight against gravity. They have no need of a large amount of stiffening material in their stems, since they are supported by water. The lack of tough, woody cells is an advantage underwater, since flexible stems can sway with the currents, instead of snapping.

The underwater leaves of plants that grow in strong currents, such as water buttercups, are often slender and branched, giving minimum resistance to the flow of water. They form a network that traps as much sunlight as possible as it filters down through the water. Long, slender, branched leaves that trail in the water also provide a large surface area for the absorption of food matter, oxygen and carbon dioxide. Some

LEFT The leaves of the giant Victoria water lily, *Victoria amazonica* of Amazonia, can grow the more than six feet across. Underneath they are supported and strengthened by thick ribs and protected by sharp spines, which probably discourage herbivores from grazing on them. Their upturned edges may prevent water pouring over and sinking them when animals stand on their broad expanse.

plant species have broad, flat underwater leaves, but these can be a disadvantage since they cannot photosynthesise efficiently if they become covered in silt or mud.

Flea-eating plants

The bladderwort has developed an unusual leaf shape. Like some other species of underwater plant, the bladderwort is carnivorous. It commonly lives in waters that lack nitrogen, and so draws the nitrogen it requires from the bodies of tiny animals (such as water fleas), which it traps in sac-like bladders formed from modified leaves. If a passing flea touches the sensitive hairs at the mouth of the bladder, a trapdoor opens and the flea is sucked into the sac, where it is digested.

Aquatic animals

In places where the current is slow, the mass of submerged vegetation creates a dense underwater forest. Here, herbivorous invertebrates move around grazing on the plants, while predatory animals, such as dragonfly and damselfly larvae, hunt the herbivores.

On the water surface, animals use the floating plants for a variety of purposes. Dragonflies take advantage of lily leaves as landing pads, while birds, such as the jacana or lily-trotter, hop from leaf to leaf as they search for food. Down in the weeds, herbivorous and carnivorous fishes lurk. Large predators, such as pike, hang motionless among the stems, waiting for unwary victims to pass by. Animals that live on land, such as otters and crocodiles, also enter the water to feed. The more plant life a river supports, the more habitats it provides for animals, creating a rich and fascinating environment.

ABOVE Water voles live in freshwater habitats, occurring along well-vegetated banks. They do not appear to have any special adaptations for swimming (such as webbed feet), but they have sharp, strong teeth to deal with the tough waterside plants on which they feed.

RIGHT Freshwater lakes in mountainous regions do not have a wide variety of aquatic life, since the surrounding rock is often poor in calcium and other soluble minerals. As a result, not enough nutrients are washed into the lakes to provide nourishing habitats.

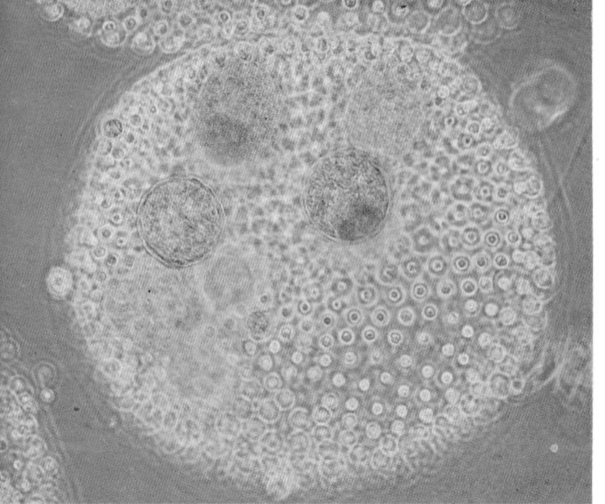

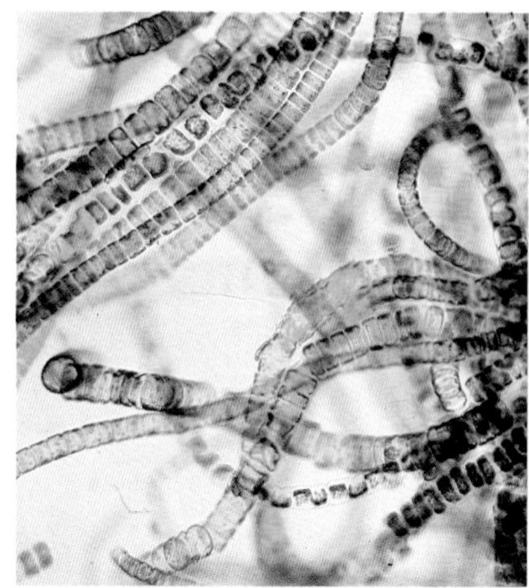

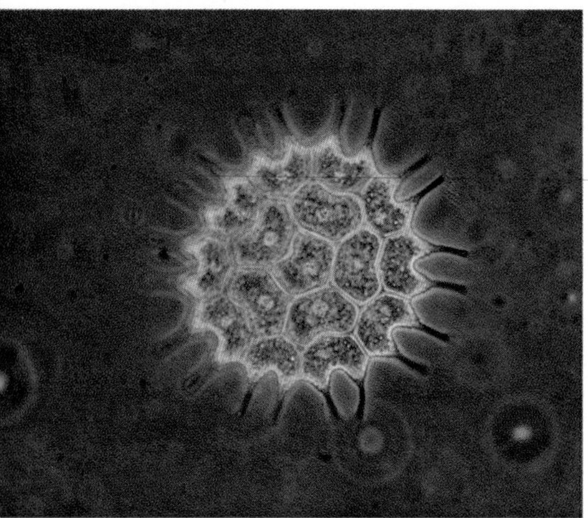

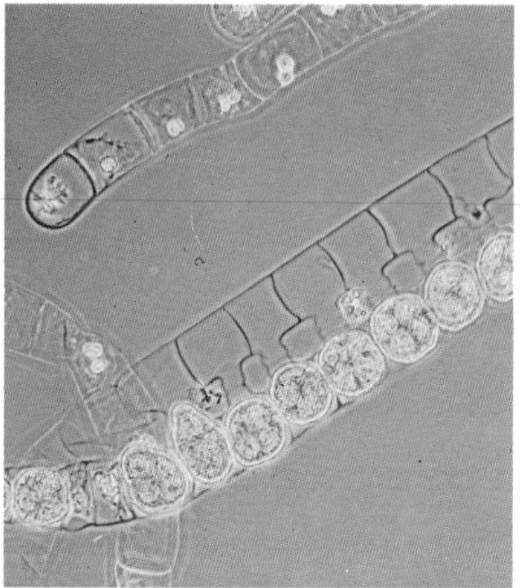

ABOVE Algae are a vital source of food for many freshwater animals, and form the foundation of aquatic foodwebs. They occur in the upper layers of still and slow-moving water where enough sunlight penetrates for them to carry out photosynthesis. The algae species include: the colonial *Volvox* (top left); *Pediastrum* (top right); *Zygnema* (bottom left); and *Ulothrix zonata* (bottom right).

Riverbank life

As the river flows downstream, and the habitat changes, the wildlife along the riverbank changes too. At first, the bank may be shaded by trees, or open and grassy. But where the river cuts into the bank on the outside of a bend, low cliffs develop, creating ideal homes for hole-nesting birds such as kingfishers and sand martins.

Water voles burrow into the bank, and adult insects, such as dragonflies and stoneflies, rest among the foliage or fly over the water. The insects are themselves prey to swallows that sweep low over the water during the day, and bats that hunt them at night.

The ecology of a lake

The main difference between the ecological conditions of a river and a lake lies in the water flow. In the river, many animals depend upon a constant supply of food being carried to them on the current. There is no such supply of food in a lake, where the waters are relatively still. Water does pass through lakes, but it does so a great deal more slowly than along even the most sluggish of rivers, and this has important consequences for the resident wildlife.

Unlike a flowing river, where oxygen dissolves to a greater or lesser degree throughout the water, deep lakes have a smaller supply of oxygen, and most of it occurs near the surface. Light, too, can only penetrate to a certain depth in a lake, depending on the clarity of the water. The limited availability of oxygen and light in a deep lake means that wildlife is most abundant at the surface.

Life at the top

The microscopic animals and plants – or plankton – that are the basis of life in the lakes, live in the surface layers of the water. Unable to survive in the fastflowing rivers (where they cannot swim against the current), plankton thrive in lakes where they can drift about. The animal plankton, comprising water fleas and other minute organisms, feed on the plant plankton. Fishes, in turn, feed on the animal plankton, and a food chain results, with powerful predators, such as pike, ospreys, crocodiles and otters, at the top.

Depth and diversity

Different types of lakes support different animal and plant species. Mountain lakes are often the poorest in life forms. Very little detritus washes into them and, although this keeps their waters clear, it starves the lakes of sufficient nutrients to support a wide variety of plants and animals.

Conversely, lakes that receive water from streams or mature rivers support more life. The shallowest lakes often have the richest flora

ABOVE The South American lungfish is an eel-like species that grows to a length of about 5 feet. It is well-adapted to swamps and the weedy margins of rivers and lakes, where oxygen levels in the warm, still waters are often low. It obtains oxygen through a pair of lungs, as well as through its gills, coming to the surface frequently to gulp air. In times of drought, it burrows into the mud, where it stays until the rains return.

and fauna, since the sunlight can penetrate to the lake bed, encouraging the growth of photosynthesizing plants. Jungles of aquatic plants growing beneath the surface of shallow lakes sway to and fro in the gentle currents.

The transience of lakes

The life of a lake is a limited one. However permanent a lake may seem to us, it is doomed to eventual extinction. The processes that lead to its end occur almost from the moment the lake has formed. At first, the nutrients in the lake increase. Vital minerals and gases, such as nitrogen and phosphorus, gradually build up as detritus is carried to the lake by rivers and streams. Added to this are the remains of plants and animals that die and fall to the bottom of the lake.

In response to the increase in nutrients, the lake's plants and animals multiply. When they die, they produce large quantities of decaying matter in the water. The effect of this is to reduce the amount of oxygen available in the water. As the quantity of organic material in the lake increases, the water becomes cloudier, preventing light from reaching the plants below. Sometimes mats of algae form on the surface, shunting out vital light and preventing other species from growing.

The detritus and sediment brought in by the streams build up on the lake bottom. At the water's edge, plants trap more material, and their own dead remains add to it. Slowly, the lake fills up with organic remains and sediment, and it begins to contract. As the plants at the lake edge encroach on the open water, the lake develops into a wetland, swamp or marsh. Finally, it will turn into dry land.

Breathing in water

All aquatic animals need a supply of oxygen, whether they live in rivers or lakes. Truly aquatic animals, such as fishes, obtain their oxygen from the water itself. The gills of a fish are well-supplied with blood, and when the water passes over the gills, oxygen transfers into the

Fish That Cope With Drought

In parts of tropical Africa and South America, where there are distinct wet and dry seasons, the killifish has evolved to deal with the climatic extremes. The adults live in habitats that often dry up, such as riverside pools and swamps, and as a result, they live for just one year. The eggs, however, are designed to survive long stretches of drought.

The killifish breeds during the wet season, producing tough-skinned eggs that it lays in the mud. Slowly but surely, as the dry season wears on, the water retreats and disappears and the adult killifishes die. But the eggs survive, remaining dormant in the mud, their development temporarily suspended.

Delayed hatching

When the rains return, the killifish's home becomes habitable once again. However, instead of hatching immediately, the eggs take eight to 10 weeks to develop before they release the young. As a result of the delay, faster-developing small animals, such as insect larvae and animal plankton, are already abundant, and can provide the fish with plenty of food.

The killifish take six to eight weeks to reach their full size – about three inches long. They are then ready to breed, and the cycle begins again.

RIGHT The 26 species of killifishes that live in the tropics of South America and Africa are sometimes called annual fishes, since their life-cycle, from birth to mating and death, usually occurs within the space of a single year. In the dry season, the females bury their eggs, which wait until the rains return before hatching.

blood. Some invertebrates, such as the larvae of some insects and all amphibians, also have gills. Invertebrates that lack gills, such as flatworms and leeches, obtain oxygen by absorbing it through the surface of their bodies.

Semi-aquatic animals, such as water beetles, take their oxygen supply into the water with them when they dive, carrying it under their wing cases. Mosquito larvae and some snails come to the surface to take oxygen directly from the air.

ABOVE The red-necked grebe, a species found in lakes, ponds and estuaries, builds its nest on open water, out of the reach of many land predators. Grebes feed on aquatic animals, and they swim powerfully underwater. Their bodies are well-suited to an aquatic life: their broad, lobed toes serve as paddles, their feet are set well back on their bodies for better propulsion, and their plumage is warm and waterproof.

Seasonal changes

The conditions of rivers and lakes never stay the same all year round. When heavy rain falls, rivers carry more water and flow faster, often raging between their banks. If the banks cannot contain the rising water, flooding results. As the water spills onto the surrounding land, the aquatic wildlife is given unexpected opportunities to range further afield.

Many rivers, especially smaller ones, flow for only part of the year. In summer they may dry up, the river beds becoming homes for lizards and snakes, rather than fishes. Some fishes have adapted to such drought in a variety of ways. The killifish's eggs, for example, survive in dried mud until the water returns (see box on page 374). During the dry season, the Australian lungfish retreats to shallow pools left in the drying-up river bed; to breathe, the lungfish rises to the surface, where it gulps in air that passes to its single lung. In this way, the lungfish can survive long periods of drought, as long as its body remains covered by some water or mud.

Lakes and their seasons

Like rivers, the amount of water that lakes hold varies over the course of the year. If the lake is large, it is unlikely to dry up completely, even in the harshest of droughts.

Lakes in temperate regions attract large numbers of birds. In springtime, they provide breeding habitats for many species of water birds, such as grebes, which nest at the lake's edge but move to the open water for courtship and feeding. In late summer and early fall, lakes support large flocks of migrating ducks, which stop off to undergo their annual molt. At this time, the ducks may be flightless and particularly vulnerable to predators, so they take refuge out on the open water.

A winter refuge

In winter, the lake may freeze. But, fortunately for the resident aquatic wildlife, water freezes from the top down, enabling fishes to retreat to the deeper levels during icy weather. Animals that depend on open water, especially birds, have to move elsewhere. Many of them migrate to warmer areas or to wetlands near the sea, where the water, carrying at least some salt, is less likely to freeze.

When temperate lakes do not freeze, they may support more life in winter than at any other time of the year. Birds such as ducks and gulls congregate on the water to take advantage of the food resources and the protection from land-based predators afforded by open water.

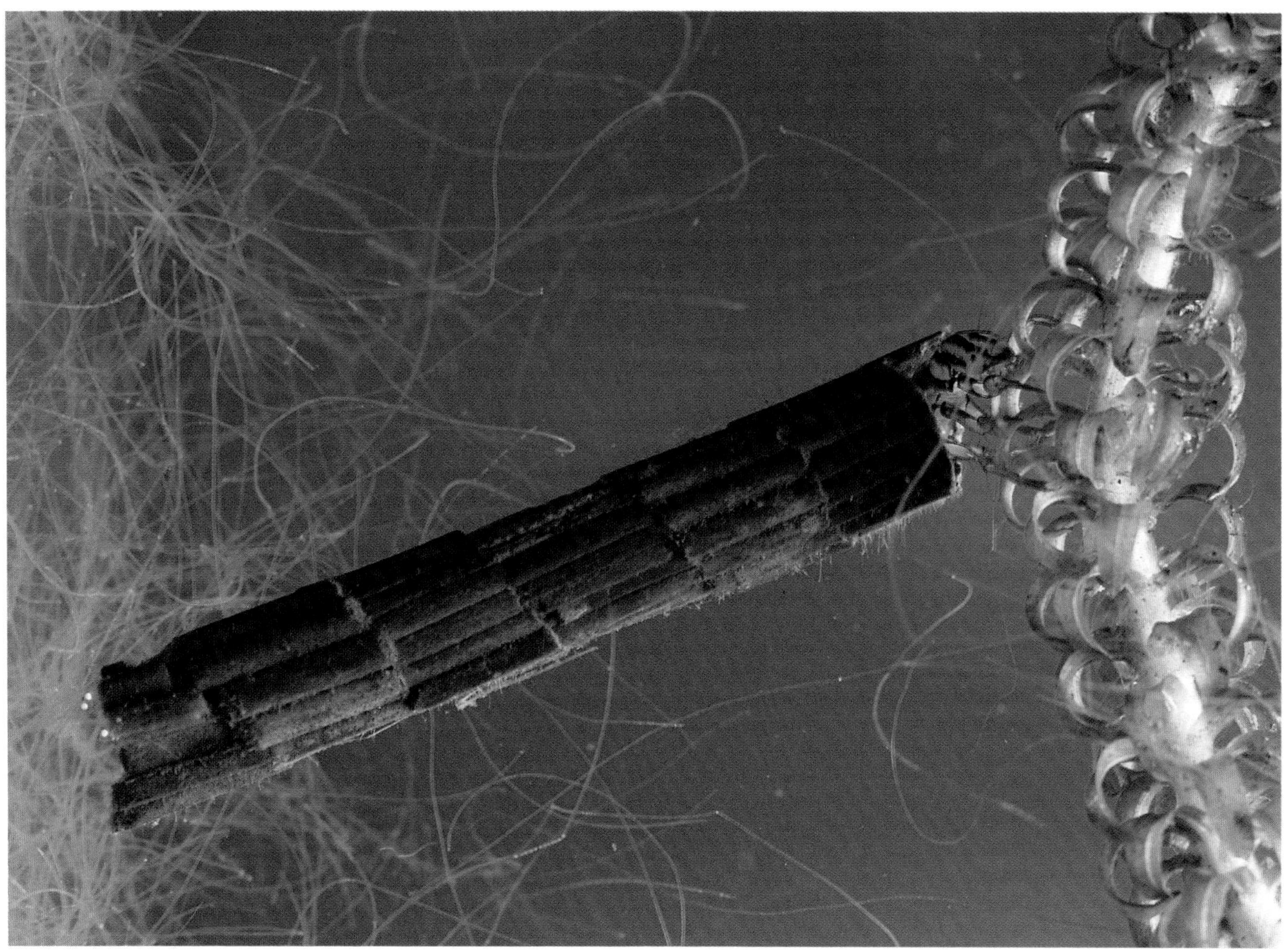

ABOVE Caddis fly larvae live in freshwater streams and breathe through external gills. Some species protect their bodies by building portable, open-ended cases of sticks (such as the one seen here), from which they partially emerge to feed on aquatic plants.

LEFT The Porcellizzo Valley in Lombardy, Italy, is typical of the early stages of a river; the course is rocky and turbulent, with many shallow stretches. Wildlife must adapt to these hazardous conditions in order to survive. Those caddis fly larvae that do not live in portable cases, for example, spin webs of silk that hold their bodies in place and trap food.

Life in European rivers

From the familiar trout to the rare Pyrenean desman (an insectivorous rodent), the rivers of Europe support a wide variety of plants and animals. The most typical creatures of the upper reaches of European rivers are the caddisfly larvae. The larvae of some species of caddisfly spin webs between the rocks on the riverbed to trap small invertebrates as they are carried downstream. Blackfly larvae are also typical of upper rivers. They attach themselves to rocks, using fan-like appendages to filter food from the water.

The inhabitants of the upper rivers have a variety of adaptations that enable them to cope with the turbulent and fast-flowing water. Small fishes, such as bullheads and stone loaches, have broad, flattened bodies with streamlined profiles; the current passes smoothly over them, rather than sweeping them away. Mayfly larvae are broadheaded and flat. They graze on the algae that grows on the rock, taking advantage of the thin layer of still water directly above the surface of the rocks. Leeches are also flat. They attach themselves to rocks, and to the fishes on which they prey, by means of suckers.

The trout

The brown trout occurs in moderately deep water. Instead of hiding among the rocks to avoid the current, it uses its powerful swimming ability to hold its position against the flow. Like all freshwater fishes, however, it must keep some power in reserve. If it is disturbed or pursued by a predator, it needs

ABOVE Trout are powerful, lithe, fast-swimming fish that can move at speeds of over 8 feet per second. They are carnivorous, feeding on aquatic invertebrates, although large specimens will eat other fishes. Trout catch invertebrates both in the water and at its surface, and will even break out of the water to reach prey.

the strength to escape. A streamlined fish, it can swim fast against the current when it needs to.

At one time, the brown trout that haunts upland streams was thought to be a different species from the larger lake trout and the powerful, silver sea trout. But it is now known that they are all members of the same species, and the differences between them are simply the result of their environment. The rich food supply in the sea enables the sea trout to grow larger, and the colors of the fish in the different locations are a result of the trout's ability to change color to match its surroundings.

Brown trout migrate upstream to their breeding grounds. River and lake trout do not travel far upstream from their homes, but the silver sea trout comes all the way from the sea. Trout spawn where there is a gravel bottom to the river. In the late fall they gather at such places, sometimes where the water is little deeper than the height of their bodies. The female trout makes a small hollow in the stream bed with her tail, and lays her eggs. The male releases his milky sperm over the eggs to fertilize them. The female then covers the eggs with gravel, and the adult fish set off downriver, leaving the eggs to hatch the following spring.

Initially the young hide among the stones, taking nourishment from their yolk-sacs. After a short period of development, they swim free and feed on small invertebrates. They themselves fall prey to other fishes and to larger predatory invertebrates, such as dragonfly larvae and water beetles.

The adult trout feed on invertebrates, including those that appear at the surface of the water, such as mayflies. Mayflies are

unusual in the insect world in that they molt twice between the full-grown nymph stage and the adult stage. When a nymph is fully grown, it climbs out of the water onto a plant, or floats to the water's surface. There its skin splits to reveal a dull-colored adult which flies up onto nearby vegetation. A few hours later, the insect molts again, shedding a thin body membrane that reveals a more brightly colored insect.

Mock mayflies

Trout attack both emerging adult mayflies and adults that fall back onto the water after their brief aerial mating activity is over. Anglers attempt to catch trout by using fishing flies that imitate the different stages of the insects' life-cycle.

Interest in freshwater fishes, both for sport and for food, has greatly influenced the ecology of many European rivers. Fishes of many types, including trout, have been introduced into places where they do not naturally occur. Carp, for example, are raised a long way from the sea in some European rivers, where sea fishes are not readily available for eating.

Marine origins

The freshwater crayfish is a large, invertebrate predator that inhabits the upper reaches of the river. Resembling a small lobster, it hunts at night, preferring clear, clean streams.

The crayfish (of which there are several hundred species worldwide) are descended from marine forms, and have inherited a concentration of salts in their body fluids that is higher than in the freshwater around them. The internal fluids of most marine animals contain the same concentration of salts as the surrounding water, preventing the salt water drawing fluid out of the animal by osmosis. In the freshwater crayfish, there is not the same balance of salts, and so freshwater passes into the crayfish's body by osmosis. To control this influx, the crayfish has evolved a waste system that produces dilute urine to expel the excess water.

Predatory moles

One of the less common predators of European upland streams is the Pyrenean desman. A member of the mole family, the Pyrenean desman is unique to the Pyrenees. It has webbed toes and waterproof fur, as well as nostrils and ears that it can close underwater. It feeds mainly at night on aquatic invertebrates. The Pyrenean desman and its close relative, the Russian desman, are threatened in the wild. In the Pyrenees, many mountain streams have been dammed, destroying the desman's habitat, while the larger Russian desman has been hunted for its fur.

ABOVE The dipper inhabits fast-running stretches of water, where it feeds on aquatic invertebrates, especially insect larvae. Its plumage is dense and waterproof, and its feet are strong, enabling it to walk along the bottom of turbulent rivers and streams as it searches for food. It frequently perches on the rocks that rise above the torrent to get a better view of its territory.

The dipper commonly occurs on the upper reaches of European rivers. It is unusual for a passerine (or perching bird) like the wrens, warblers and finches, to be so closely associated with water.

The dipper hunts and feeds in the water. Lacking webbed feet, it uses its wings to propel it down to the bottom of the river, where it walks along the riverbed searching for invertebrates. The dipper can stay underwater for as long as 30 seconds. It has a habit (from which it takes its name) of bobbing up and down on exposed rocks between bouts of hunting. Dippers can also feed under ice, an adaptation that enables them to remain in their territory throughout the year.

LEFT **The Eurasian kingfisher is stocky and streamlined, and dives deep into water to catch fish. It usually hunts from a perch, flying rapidly down into the water when it spots a potential victim. Once underwater, it grasps the fish in its powerful beak, then flies back to its perch with its prey, which it swallows head first; this technique prevents the fish's spines or fin-rays sticking in the kingfisher's throat.**

The dipper builds its ball-shaped nest in a concealed place among tree roots or rocks. It will even nest behind a waterfall, flying through the water to reach its nest. The dipper's eggs, like those of many birds that breed in closed nests or in holes, are white – an adaptation that helps the parent birds to see them in the dim light inside the nest.

High pitched calls

When a spring thaw increases the volume of water in a river, or when the river passes over a water-fall, the flow of water becomes a noisy, rushing torrent. In order to communicate in such a noisy environment, various birds, such as the dipper, the gray wagtail (in Europe), the forktail (in South-east Asia), the plumbeous redstart and the white-capped river chat (in the Himalayas), have simple, piercing, high-pitched calls. These loud, clear calls, which are used either as an alarm or as a way of making contact with another member of the same species, can easily be heard above the rushing waters. The birds frequently bob up and down on the rocks in a display to potential mates since complex courtship songs would be useless in the roar of a torrent.

Kingfishers in winter

Although some kingfishers breed in the uplands, they are much more typical of the weedy reaches of lowland rivers. Kingfishers feed on small fishes such as three-spined sticklebacks. They perch or hover above the water, watching for their prey, and dive down into the water to snap up their victim. The kingfisher, like all birds that are associated with freshwater in temperate latitudes, has to overcome the difficulties presented by hard winters when the water may freeze. While the dipper often stays put and can survive as long as it can get under the ice to feed, the kingfisher is unable to penetrate the ice in search of its food. It has to hunt elsewhere, and in hard winters the kingfisher moves to lowland rivers and coastal waters.

ABOVE Whirligigs are small carnivorous beetles that spin round in circles on the surface of still or slow-moving freshwater. Their eyes are divided laterally – they use the lower halves to search for prey below the surface of the water, while the upper halves keep watch above the surface. Whirligigs have hairy, oar-shaped legs that enable them to swim well.

Although the kingfisher moves to ice-free water in the winter, the harsher months can be disastrous for it since competition between individuals increases. In extremely harsh conditions, when lowland waters freeze, many kingfishers die. After a cold winter in 1962-3, the kingfisher population of Wales was estimated to have fallen by as much as 85 per cent.

Life in weedy waters

During summer, when the weeds increase in the slower reaches of the river, the water is inhabited by a wide range of animals including water snails, flatworms, freshwater bivalves, worms, leeches, crustaceans and the larvae of mayflies, stoneflies, water beetles and dragonflies.

Sand and silt often get caught up in the current, making the water abrasive. To protect themselves against damage, the larvae of mayflies and stoneflies have hairy bodies. The caddis fly larva builds a casing of pebbles around itself. Leaving its head and front legs protruding from the casing, it can scavenge for scraps of food in relative safety.

Crustaceans that inhabit freshwater environments include water slaters and freshwater shrimps. Water slaters have flattened bodies and resemble aquatic woodlice. They scavenge for food on the bottom of the river or among the aquatic vegetation.

Water snails are mostly herbivores that feed on the aquatic plants. However, some species, such as the great pond snail, are partly carnivorous and will feed on injured fishes as well as plants.

Worms are an important food source for fishes, and become increasingly abundant in muddy riverbeds into which they can burrow. The bivalves, such as zebra mussels and swan mussels that live in the mud of the riverbed, lie buried with two siphons protruding from the shell. One, the inhalant siphon, draws water in, and the other, the exhalant siphon, passes it out again. Oxygen and food particles are taken from the water as it passes through, and waste products are expelled.

The young of many species of bivalve live as parasites on the fins and tails of fishes before they drop down to the bottom to start their mud-dwelling lives, where they become an important source of food for waterfowl and mammals.

The buffalo gnats

In European rivers, six or seven species of buffalo gnats (or blackflies) occur downstream from the river's source. The species *Simulium lineatum* and *S. equinum* (the horse buffalo gnat) are both common in northern Europe. Because they do not need the same breeding conditions, their ranges do not overlap. *S. equinum* lays its eggs on plants that trail in the water, and so it lives in turbulent or quiet tributaries with plenty of bankside vegetation. The eggs of *S. lineatum* must fall to the bottom of the river to develop, and so this species lays them on quieter stretches of the main river, where they are less likely to be carried away by the current.

Small but fierce

The abundance of small invertebrates in the river attracts larger, carnivorous invertebrates such as dragonflies and damselflies. Some of these predators are extremely

Fish Raised In A Mussel

The bitterling is a common fish of muddy, weedy, European freshwaters, and is notable for the unusual method it has of taking care of its young: it uses a freshwater mussel, often a swan mussel, as a nursery.

During the breeding season, which lasts from April until August, the female bitterling develops a pointed, tubular organ known as an ovipositor that measures about two inches long. After finding a freshwater mussel, the female places her ovipositor into the bivalve's inhalant siphon – the tube that draws in food and oxygen – and lays up to 100 eggs.

The male bitterling then swims over the inhalant siphon and discharges his sperm; this is sucked into the mussel's gill cavity where the eggs are lying, and fertilization takes place. The male will lure more than one female to the mussel, and since several males often use the same mussel, the bivalve may eventually contain several hundred eggs, all at different stages of development.

A mussel's protection

The mussel is a good nursery for the eggs and young fish. They are kept well-supplied with oxygen, and are safe from predators. They are even protected from drought, because the mussel will move to deeper water if the water dries up in hot weather. About a month after hatching, the young bitterlings swim out of the mussel, leaving it completely unharmed. As with many such close associations in nature, both the mussel and the bitterling benefit from the relationship. The mussel's larvae, produced during the bitterling's breeding season, often depend on the fish for their survival. They will attach themselves to a bitterling's gills and fins, and feed on the fish's body fluids for several weeks until they drop off and begin their lives as independent adult mussels.

BELOW A male Japanese bitterling (*Rhodeus ocellatus*) guides his pregnant female partner towards the freshwater mussel in which she will lay her eggs. She trails her long ovipositor, which she uses to insert into the mussel and through which she lays the eggs.

fierce, and attack virtually anything of suitable size that comes within range of their mouthparts.

Dragonfly and damselfly larvae are as predatory as the adults. They lurk among vegetation or at the bottom of the water, waiting for prey to pass by. When an unsuspecting invertebrate comes within range, the larva shoots out its "mask" – a modified labium (or lower mouthpart) whose hinged segments fold up in front of its face when at rest. Hooks at the end of the mask grab the prey, and the mask then brings the prey back within reach of the strong jaws.

Water beetle larvae are equally ferocious predators. The larva of the great silver beetle feeds on water snails. The adult, however, is a vegetarian and feeds on aquatic plants. The larva and adult of the great diving beetle are both carnivorous, catching tadpoles, other insect larvae and even small fishes in their powerful mouthparts.

Zones of convenience

Temperate rivers are often divided into four zones that are linked with particular fishes: the trout zone, the grayling zone, the barbel zone and the bream zone. These zones reflect ecological differences along the course of the river, especially its oxygen content.

In the trout zone, the water is usually quite shallow, fast-moving and cool, and it has the highest oxygen content of all four zones. Apart from trout, other fishes that occur in this zone include stone loaches and bullheads.

The grayling is typical of the cool, clear, clean waters that occur further downstream, although it also occurs in lakes. It feeds on a variety of invertebrates, including terrestrial insects that fall into the water. Other fishes that inhabit the grayling zone include riffle minnows, chubs, nase and gudgeon.

The barbel lives in shoals in deeper, fast-running waters. It feeds mainly at night, poking around the sandy riverbed and turning over rocks in its search for bottom-

ABOVE The larvae of dragonflies, such as the hawkers, are among the most fearsome of all freshwater predators. They detect their prey with their sensitive eyes and antennae, and catch it with their unique extending mouthparts, which they shoot out to seize the victim in 1/25,000 of a second. They will tackle fish, such as sticklebacks, which are virtually as large as themselves, as well as snails, limpets, tadpoles and crustaceans.

dwelling invertebrates. The barbels (fleshy filaments) around its mouth are well-equipped with taste buds, enabling the fish to test food before eating it. The dace is also common to the barbel zone.

The bream inhabits stretches of slow, muddy rivers that are near to the sea. It feeds on mud-dwelling invertebrates, such as worms and midge larvae, sucking them up through its mouth, which extends to form a suction tube. When the bream is feeding, it adopts an almost vertical posture in the water, seeming to stand on its head over the bottom.

The bream zone often has a low oxygen content, since there is little surface movement to increase the amount of oxygen dissolved in the water. The animals and plants use much of the available oxygen for respiration, while decomposition also requires a large amount of oxygen. Fishes that inhabit this environment include bleaks, bitterlings, ruffes and catfishes.

Parents great and small

The three-spined stickleback is common in streams, ponds and lakes throughout Europe. Like many fishes, it displays complex courtship and mating behavior. In spring, the dull, greenish-brown male changes his color, becoming bright red underneath and blue on his back. He builds a nest of plant material, and glues it together with a substance secreted from his kidney. He then entices a female to lay eggs in the nest, and fertilizes them. After chasing the female away, the male looks after the eggs himself, regularly fanning them with his pectoral fins to keep them well-supplied with oxygenated water until they hatch 10–14 days later.

Parental care

Different species of fish provide different degrees of parental care for their offspring. The three-spined stickleback takes great care of a relatively small number of eggs, usually less than 100. Other fishes, such as the trout, produce large numbers of eggs and then abandon them, relying on the survival of a few to guarantee the continuation of the species.

The beluga, a huge sturgeon which has been known to weigh more than 500 lbs, lives mainly in the sea, but swims up the great rivers of Europe, such as the Danube, the Volga and the Don, to spawn. The female produces several million eggs, and the young feed on bottom-dwelling invertebrates before moving on to a diet of fish.

ABOVE Pike are fierce freshwater predators that measure up to 5 ft long. There are few freshwater creatures they will not attack; waterfowl and water voles, as well as fishes as large as salmon, all fall victim. Pike have long, slender bodies, with their fins concentrated at the rear end; this allows them to shoot forward very quickly over short distances to catch their prey. They hunt by sight, and usually live in clear, but weedy water, where their stripes camouflage them in the vegetation.

ABOVE Otters chase their prey underwater. Their streamlined bodies and webbed feet enable them to swim quickly, twisting and turning in pursuit of fishes which they catch in their jaws. When they land their prey, otters handle it with their highly developed, sensitive paws and crush the fish bones easily with their strong teeth.

The beluga is an important food fish for humans who eat both the flesh and the roe, which is familiar as caviare. But as with many commercial species, the beluga has been overfished during the 20th century, and their numbers have declined drastically.

Riverside life

The European water vole occurs wherever there is plenty of bankside vegetation – grasses and herbs – to feed on. It swims well, paddling along on the surface using all four feet, and only dives if threatened. It is mainly active during the day.

Insects buzz, drift and swarm over the water both day and night. During the day, swallows, martins and swifts snap them up; at night, bats replace the birds as insect hunters. Both the pipistrelle and Daubenton's bat feed along stretches of European rivers. Daubenton's bat has a graceful flight and swoops low over the water. As it finishes its pass at the end of a stretch of river, it curls up into the air and swings round to return along its route.

Lake invertebrates

Most of the animal plankton that occur in lakes are minute or microscopic crustaceans. Some of them – water fleas of the genus *Daphnia* – measure tenths of an inch long and are visible to the naked eye. Water fleas are an important source of food for fishes such as roach and young perch. They are also eaten by the tiny, tentacled hydra – a freshwater relative of the jellyfish. The hydra anchors itself to vegetation, and waves its tentacles about in the water. When a water flea strays within reach, the tentacles trap it, stinging and paralyzing the flea with specialized cells called nematocysts.

Many animals take advantage of the surface tension of the water and live at the lake surface. They

ABOVE Water shrews, which are active during both day and night, carry their food ashore to eat. Although small, they are ferocious predators and will attack prey larger than themselves. They subdue their victims, at least partly, with their venomous bite; the poison they produce affects the nervous system.

generally have slender bodies and water-repellent hairs at the ends of their legs, enabling them to skim across the water quickly and easily.

The six-legged pond skater, for example, feeds mainly on insects that have fallen onto the lake, hunting for them among the duckweed and water lilies. It moves around on its two rear pairs of legs, using the front pair to catch its food. It is sensitive to disturbances in the surface tension, and will rush to investigate them. The pond skater belongs to an order of insects with piercing and sucking mouthparts. It kills its prey by injecting it with a toxin and then sucks out its victim's body fluids.

Water scorpions and mosquito larvae hang from the undersurface of the water, and breathe the air above. Mosquito larvae feed on algae, but the water scorpion is a predator. It hangs upside down waiting for tadpoles, small fishes and insects to come within range, and grabs them with its front legs. It breathes by means of a long, respiratory tube that stretches from its hind end to the air above the surface.

Predatory pike

The abundant plant and animal life of a lake provides a rich source of food for a wide variety of fishes. Some fishes feed on animal plankton, while others feed on plants. Roach, for instance, eat both plant material and the animals concealed among it.

Fishes are themselves food for fierce predators such as the pike, which lurks among weedy vegetation. It surges out of its hiding place to grab any suitable prey that swims by, and will even eat chicks, dragging them down from the lake surface. To prevent predation by pike, great crested grebes carry their young on their backs as they swim around in the breeding season.

Pike are fairly common in urban freshwaters, where they thrive in the absence of other vertebrate predators such as otters. Otters are shy creatures that only inhabit undisturbed areas where the water is clean. They are fast, powerful underwater swimmers that feed exclusively on fishes. They have webbed feet, waterproof fur and ears that can be closed underwater.

The water shrew is a smaller predator of European rivers and lakes. Its hair-fringed feet give it an agility underwater, where it hunts for small fishes and aquatic insects. The water shrew's coat is specially adapted to trap air, enabling the animal to float on the surface.

Hollow bones, webbed feet

Herons and ducks are common lake birds. The gray heron occurs in the shallows of European lakes, where it wades slowly from one spot to another in search of fishes. When

it sees one, the gray heron thrusts its head below the surface of the water and grabs it. Ducks have buoyant bodies with hollow bones that they fill with air directly from their lungs. They have webbed feet and powerful legs, and are strong swimmers. Their thick plumage is tightly arranged, making it extremely waterproof.

Waterbirds use a variety of feeding methods to avoid competition with other species, and to fully exploit the available resources. Diving ducks, such as the pochard and the tufted duck, dive underwater to feed on shellfish, worms, insects and small fishes, as well as some plant matter. The black-throated diver is a superb swimmer, and may hunt fish and aquatic invertebrates at great depths; the great crested grebe catches fishes that swim in open water, relatively close to the surface.

Dabbling ducks, such as shovelers and teals, obtain the plant matter that they eat by probing their heads underwater (dabbling), or by upending themselves to reach food just below the surface. The mute swan uses its long neck to reach food – mostly plants, but also some frogs, mollusks and insects – in deep water.

The mallard combines dabbling and diving with grazing arable fields and grubbing for roots and invertebrates. Mallard occur over much of western Europe, although in the far north they are only summer visitors. Since they can take off almost vertically from a confined space, they can nest on even the tiniest pools. In contrast, great crested grebes only breed near more open waters, since they need a good run across the water to take off.

ABOVE The black-crowned night heron is one of the most widespread of the herons, occurring in wet places in many parts of the world (the bird shown here is from Hawaii). Although they feed mainly at night, using their large eyes to detect prey in the dark, they also feed during the daylight hours, especially in the breeding season. They stalk fishes and small reptiles, catching them with a darting thrust of the head.

The osprey hunts for fishes from the air, swooping down to the water to grab its prey. The osprey is not the only large carnivore of the lake. Because of the abundance of water birds, several species of harrier, such as the marsh harrier, occupy lake habitats. The marsh harrier is a magnificent bird of prey that

hunts by gliding silently over the lake, plummeting down to take its victim, such as a grebe, coot or young swan.

Migrating birds

In summer, many waterbirds breed on the northern European lakes, taking advantage of the rich supply of plant and animal life. When the cold weather arrives, the lakes often freeze, and the water birds return to warmer areas. If the lakes do not freeze, they attract huge flocks of wildfowl. There are often far more birds on the water in winter than there are in summer. Typical northern breeders in Europe include pintails, teals and wigeons, golden-eyes, smews and goosanders. They are joined in winter by huge flocks of gulls that roost on the open water.

LEFT The mute swan is Europe's most common breeding swan. It feeds on aquatic vegetation, which it reaches by dipping its long neck into the water or by up-ending its body so that it can stretch down to the bottom. The bird also grazes on vegetation that grows at the water's edge.

Buffalo gnats in Africa

Buffalo gnat (or blackfly) larvae are found in fast-running streams in many parts of the world. Streams in Africa often have large populations of these small insect larvae, which can survive in the fiercest torrents. They live in dense groups on rocks on the bottom of the stream, their bodies pointing in the direction of the flow.

To prevent being swept away by the current, buffalo gnat larvae anchor themselves to rocks and stones with silken threads. If they are carried off, they can reel themselves back along their threads in a similar way to a spider hanging from a web or a caterpillar dangling from a leaf.

Adult female buffalo gnats bite birds and mammals, including humans, to take the protein-rich blood they need for the development of their eggs. In Africa and South America, buffalo gnats transmit river blindness disease.

ABOVE The osprey catches fishes that swim near the surface of the water. It flies low over the water until it spots a fish wish its sharp eyesight. The osprey then plunges towards the water, stretching its feet out just before it hits the surface. Its talons are long and sharp, and its feet are scaly, giving the osprey a good grip on its slippery prey.

A GUIDE IN THE DARK

Fishes that inhabit muddy, dark waters where the visibility is low, and fishes that are mainly active at night, have evolved a range of effective sensory devices to enable them to navigate and hunt for prey.

Catfishes have up to eight pairs of slender, sensitive whiskers (barbels) that can pick up even the slightest disturbances in the water. They live in rivers or deep lakes where water plants are abundant, spending the day under overhanging banks or on the mud in deep water. They forage in the mud with their barbels in search of small invertebrates. Some Asian species of catfish have barbels that are almost half as long as their bodies.

Electrical impulses

Some fishes have evolved a range of electricity-generating organs for navigation and defense purposes. The electric eel, which inhabits the freshwaters of South America, emits two kinds of electric impulses. The first, and least powerful, are used in navigation. The impulses rebound off obstacles in the water, enabling the eel to judge with precision when to change direction. The second type of electric impulse is used for defense and killing prey, such as frogs. The eel can deliver a powerful electric shock of up to 600 volts (enough to stun or even kill a human being).

Elephant snouts are large, bottom-dwelling fishes that inhabit the lakes and rivers of Africa. They generate a continuous field of weak electrical discharges, creating an electrical field around themselves. Distortions in the electric field enable the elephant snouts to detect approaching prey and potential mates among the sediment of their murky habitat.

BELOW Catfishes, such as this naked catfish of Africa, are named for the whisker-like barbels that sprout from their lips and snouts. The barbels enable the fish to feel their way about the often dark and murky depths of their habitat, and to locate food. Most catfishes live in freshwater, but some occur in tropical and subtropical seas.

Freshwater giants

African rivers and lakes contain some large and impressive animals, including the Nile monitor (a species of lizard) and the hippopotamus. The Nile monitor measures six feet long, and is an excellent swimmer. The female lays about 35 eggs in holes in riverbanks or in nooks in trees that are close to the water. Although it eats mainly aquatic animals, such as fishes, mollusks and frogs, the Nile monitor sometimes raids crocodiles' nests for eggs – a dangerous activity, since crocodiles defend their nests vigorously. (The female Nile crocodile even guards her young in nursery areas for several months after they hatch.)

The hippopotamus is a land animal that has adapted to exploit the freshwater environment. It feeds mainly on land, and uses the water for shelter from the sun and for protection from predators. Hippos are sociable animals that live in groups of between about five and 15 individuals. During the day, they lie in the water and on sand banks; they are not strong swimmers, but appear almost graceful as they "walk" along the riverbed, buoyed up by their great bulk. During the evening, they leave the water to graze on grasslands near the river. Although they eat some aquatic vegetation, they live mainly on grass, and will consume as much as 130 lb in a single night's grazing.

ABOVE The hippopotamus is the largest of the animals that inhabit Africa's freshwaters. A herbivore that weighs up to 3 tons, the hippo spends most of the day resting in the shallows – its immense bulk buoyed up by the water.
PAGES 390–391 The Nile crocodile has certain adaptations in common with the hippo; both, for example, have eyes and nostrils on the tops of their heads, enabling them to both see and breathe while the rest of their bodies are submerged.

The dam-building beaver

Beavers are herbivorous mammals that are best known for their dam-building activities on North American streams (although some subspecies live in northern Europe and Asia). Using sticks, logs and

ABOVE The beaver has large, sharp incisor teeth that enable it to gnaw and fell tree trunks, which it uses for food and for building dams and lodges. When underwater, the beaver can shut its lips behind its teeth and close off its throat with the back of its tongue, enabling the animal to continue using its teeth when submerged, without choking. Other adaptations to an aquatic life include the flat, paddle-shaped tail, webbed feet, waterproof fur and streamlined body.

mud, they construct dams across the rivers to form ponds, in the center of which they build their lodges or homes (from the same materials). Placed in the middle of the water, the lodges provide effective protection from predators. With their dam-building and stream-deepening activities, beavers have a considerable effect on the land-scape and ecology of their habitats.

Other North American animals that take advantage of rivers include the American dipper and the grizzly bear. The American dipper, like its European relative, favors fast-flowing streams. It lives as far north as Alaska, where its range overlaps with that of the grizzly bear. Grizzly bears take advantage of a rich food supply when salmon run up the rivers to spawn. The bears are skilled at fishing, and catch the fishes with their claws or their teeth, taking them ashore to eat.

Saltwater lakes

Salt lakes are often formed when rivers accumulate and carry down salts from the surrounding hills into a lake. If the lake has no outlet, the salts remain, increasing in concentration as the water evaporates. Two examples of salt lakes are Lake Mono in California and Lake Natron in East Africa.

Lake Mono supports great numbers of brine shrimps and brineflies, but fishes are unable to survive in the highly saline water. Without fishes, competition for the available food is greatly reduced, and waterbirds are attracted to Lake Mono in great numbers. These include a large variety of gulls, grebes, waders and ducks – up to 800,000 birds have been recorded on the lake in a single day.

Lake Natron lies in the Great Rift Valley of Africa, between Mount Kilimanjaro and Lake Victoria. Its waters contain large quantities of soda (alkaline salts) which come from volcanic geysers. Although highly alkaline, Lake Natron supports an abundant, if restricted, range of animal and plant life, which includes algae, insect larvae and small mollusks. The land surrounding the lake is barren and sterile, and there are few predators. Such a safe, isolated position attracts large flocks of flamingos. The birds build mud-mound nests in the shallow lake and filter-feed on the rich supply of food in the water.

Unique species

Lake Baikal in the USSR is more like an inland freshwater sea than a lake. It is more than 5,000 ft deep, and holds 80 per cent of all the standing freshwater in the USSR. The lake supports about 1,700 species of animals and plants, of which more than two thirds are unique to the area. The animals include the Baikal seal, which is one of only two freshwater seals in the world. Lake Baikal also supports flightless caddisflies. Instead of flying, they use their legs and paddle-like

ABOVE Lesser and greater flamingos breed in vast numbers at Lake Nakuru in Kenya. However, the lake is not a hospitable freshwater habitat; instead, it is a soda lake, in which the water is full of alkaline salts that support mainly algae and invertebrates (such as crustaceans and mollusks) in vast quantities. The lesser flamingos, seen here, are the smallest of all flamingo species, measuring 30–50 in in height. They harvest the algae by straining the surface water through their specially adapted bills, which are equipped with a set of filtering plates.

RIGHT The Baikal seal lives only at Lake Baikal in the USSR. Although it is a freshwater seal, it shares similar aquatic adaptations with its marine relatives. It has a torpedo-like shape and large, webbed hind feet that it uses for swimming.

ABOVE The largest of the piranhas, *Pygocentrus piraya*, measures some 2 ft in length and is widespread in the lower Amazon basin. It has strong jaws, and its teeth are stout, triangular and razor sharp, allowing the fish to cut swiftly and efficiently through the flesh of its victims. The piranha's carnivorous habits have given it a fearsome reputation.
PAGES 398–399 The giant otter of Brazil is the largest of the world's otters, reaching almost six feet long and weighing up to 50 lb. Unlike other otters, it has fully webbed forefeet that enable it to swim rapidly underwater in pursuit of fishes. The last two-thirds of its tail are almost flat, giving the otter extra maneuverability.

wings to swim across the lake surface. Other unique lakeland creatures include species of grebes that live on Lakes Titicaca and Atitlan in South America.

South American riches

The tropical aquatic ecosystems contain a wide diversity of wildlife. The Amazon, for example, contains about 1,300 species of fishes, compared to the 192 species that occur throughout the whole of Europe. Other tropical areas are similarly rich in fishes: the Zaire River contains 690 species, while the rivers of Thailand have 546 species.

Among the fishes that inhabit the Amazon River, the piranha is probably the most notorious. Its fearful reputation is largely exaggerated, since the piranha usually only attacks other fish and dead or injured animals that are stranded in the water. Carnivorous fishes such as the piranha are able to find food all year round, but for many Amazon fishes, food is not so easy to come by. Many of the plant and animal plankton on which the fishes feed rely on the nutrients and dead matter in the water for their food and development. However, the rain forest that surrounds the Amazon

THE FEATHERED FISHERS

A number of birds, including darters, skimmers and fishing owls, have evolved a variety of adaptations that enable them to catch freshwater fishes. Darters live in Africa, the Americas, Asia and Australia. They are slender waterbirds that swim with only their heads and necks sticking out of the water – unlike ducks, they do not have waterproof plumage to keep them buoyant.

Darters stalk their prey slowly underwater. They catch aquatic invertebrates and fishes, which they stab with their dagger-shaped beaks. When they emerge from the water, darters often sit on branches or fence-posts, stretching their wings to let them dry.

Scissor-bill

The skimmer, which occurs around African, Asian and American rivers and coasts, has an extraordinary adaptation for catching fish. Its bill has two flattened mandibles that shut together like scissors – the shorter, upper mandible fitting into a notch on the longer, lower mandible. Skimmers fly low over the water, shearing the surface with their lower mandibles. When they hit a fish, they snap the bill shut on their victim.

Pel's fishing owl is a nocturnal predator along African rivers. It swoops down to take its prey from the water, seizing the slippery fish in its sharp talons. In South America, fishing owls are replaced by bats such as the fishing bulldog bat, which has huge feet with sharp claws and long legs. Scientists believe that the bat catches its prey by echolocation, registering where the ripples of a fish break the surface.

BELOW The buffy fish owl is a fish-eating species of South-east Asia, where it lives along the forest streams. Like other owls, it has superb eyesight, a sharp, hooked bill and sharp talons. The lower parts of its legs are bare, so that the owl can plunge its feet swiftly in and out of the water to catch prey without its plumage becoming waterlogged.

recycles its nutrients so efficiently that little reaches the river.

The Amazon, like many tropical rivers, floods during the wet season, inundating vast areas of forest. When this happens, the fish swim out over the flood plain and gorge themselves on fruit, seeds, insects and anything else that falls into the water. They eat enough food to last them through the lean period until the next flood.

Rare giant otters

Many large fish-eating predators in Amazonia feed on the enormous numbers of river fishes. The giant otters of the Amazon, for example, follow the fishes out over the flood plain, where they catch and eat them. Giant otters have suffered from habitat loss and from fur-hunting, and are now the world's rarest otters.

River dolphins and reptiles

Freshwater dolphins live in some of the world's great rivers, including the Amazon, the Ganges and the Yangtze Kiang. They feed mainly on crustaceans and fishes, especially catfish. To cope with its muddy river habitat, where visibility is low, the Amazon River dolphin, or bouto, navigates through the water and hunts using both sight and echolocation. Both the Ganges and Chinese river dolphins have poor eyesight, and rely mainly on echolocation to navigate and find their fish prey.

The Amazon is home to many species of reptile, including the matamata, a species of turtle. It has a flattened body, the skin on its neck and head hangs in short flaps, and its shell is often covered in algae. Totally camouflaged on the riverbed, it lies hidden until an unsuspecting fish passes by. Then the matamata opens its huge mouth and sucks in its prey. The flaps of skin on the matamata's chin and neck are well-supplied with nerves, and may assist the turtle in detecting prey by registering disturbances in the muddy water.

ABOVE The matamata turtle of the Amazon is one of the strangest looking turtles, with its unusually bumpy shell, frilled neck and wide, triangular head. Mainly a bottom-dweller, the matamata raises its flexible nostrils above the water to breathe, while the rest of the turtle remains submerged. As matamatas age, algae gradually accumulate on their shells giving them even better camouflage underwater.

The Amazonian manatee or seacow is a large herbivore that weighs up to a ton or more and lives permanently in the water, feeding on surface plants. The manatee is especially valuable to the Amazon river system since it clears channels clogged by floating vegetation.

The water fern weevil is another natural controller of water-borne weeds. It feeds on a South American water fern *Salvinia molesta* – one of the world's most destructive weeds. Forming a thick carpet on the surface of the river, the weed cuts off light from the water beneath, reducing its oxygen level and killing many aquatic animals. Water fern also blocks waterways so that boats cannot get through, seriously interfering with people's livelihoods. The water fern weevil eats such large quantities of fern, that in good conditions, it can double its body weight in two days.

ABOVE Even the famous canals of Venice are not free of the refuse that often spoils urban waterways. Some aquatic wildlife has been able to tolerate the disturbance and garbage that are typical of city life, but few animals can tolerate the invisible pollution of industrial wastes and agricultural runoff. The chemicals poison freshwater and kill the wildlife of many rivers and lakes.

Exploiting rivers and lakes

The use of rivers and lakes for drinking water, irrigation, transport, electricity and leisure has led to the taming of many wild rivers, especially in Europe and North America. Rivers have been dammed, their courses straightened, their flow interrupted for hydro-electric power schemes and their banks cleared. Tampering with rivers in this way has had devastating effects on the wildlife. Whole valleys have been lost to reservoirs, while so much water is taken from rivers that they often dry up. The Colorado River, for example, peters out in a sandy bed miles from its true mouth at the head of the Gulf of California in north-west Mexico – its water having been drawn off for massive irrigation and industrial purposes. While cleverly-engineered channels with smooth concrete banks control the flooding of unruly rivers, they also destroy the habitats of many animals.

The need for water has created new freshwater habitats, such as reservoirs, lakes and water-filled gravel pits. In northern Europe for example, many lowland lakes are artificial. They represent a real opportunity for the conservation of freshwater wildlife, and could even be run as nature reserves.

Human devastation

Even though we require fresh water to survive, many rivers and lakes, through neglect and carelessness, are heavily polluted. In their quest for ever greater levels of production, farmers pour large amounts of pesticides and fertilizers onto their land. Unfortunately, these chemicals do not stay on the land, but flow into rivers and streams, poisoning both the water and its wildlife.

Even recreational activities have adverse effects on the wildlife: water-skiing disrupts waterbirds and drives them from their nesting sites, while the wash of power boats often swamps the birds' nests.

Loss of habitat and excessive hunting cause immense damage to wildlife, reducing populations of creatures such as the Baikal teal, which breeds beside lakes and rivers in north-east Asia. If these and other animals, including dippers and giant otters, are to survive, many rivers and lakes must stay wild, or be treated with respect. Greater pollution will only lead to the extinction of thousands of animals and plants.

MARSHES AND SWAMPS

Supreme wetland habitats, marshes and swamps are halfway zones between dry land and open water, where bulrushes, reeds and bald cypresses form dense, jungly growth in the shallow water, and amphibians, reptiles and wading birds stalk their prey

Marshes and swamps are wetland habitats with the features of both water and land ecosystems. Because of this, they contain a particularly rich and fascinating wildlife.

Freshwater swamps and marshes are scattered across the world. They occur at the edges of streams, rivers and lakes, in the upper reaches of river deltas and in poorly-drained areas. Some are tiny, and form mere fringes around a pool, or beside a brook, whereas others, such as the Okavango in Botswana and the Pantanal in Brazil, are enormous.

The frequent occurrence of wetlands, however, does not guarantee their survival; they can be easily drained, and they are among the most fragile and threatened habitats in the world.

Worldwide wetlands

Europe's major freshwater wetlands include the Coto Donana in south-west Spain, England's Norfolk Broads and the Camargue, which forms the delta of the River Rhone in France. North American swamps and marshes include the bayous (or channels) of Louisiana and the Everglades of Florida. South America has the marshes of the Pantanal along the Paraguay River; and Africa, the Okavango flood plain in Botswana.

Swamps and marshes are not the only wetlands in the world. In the north of Europe, Asia and North America there are great sweeps

ABOVE The purple heron is a characteristic swamp-dwelling bird of Africa, Asia and southern Europe. Tall and slender, it shares with other herons adaptations that enable it to feed in shallow water. It has long, thin legs on which it stalks the reedy vegetation and open waters for food. When it spots a suitable fish, amphibian or insect, it stabs at it rapidly with its long, sharp bill. The purple heron usually nests in dense reedbeds, where its shape and coloration give effective camouflage amid the reeds.
RIGHT Swamps are often dominated by tall plants such as papyrus (foreground) and reeds, which thrive in wet conditions, and provide food and shelter for animals.
PAGES 402–403 Swampland, with its mixture of open water, wet ground and tufts of vegetation, supports a variety of creatures, from dragonflies and herons to alligators.

DRAGONFLIES ON THE MOVE

Water, whether it is a stream, river, pool, lake or wetland, is the center of a dragonfly's life. Dragonfly larvae are fully aquatic, and adult dragonflies hunt, court and lay their eggs in and around the water.

Dragonflies spend up to a year or more underwater as nymphs. Once they emerge into the open air, they will live for a far shorter period, completing their life-cycle during their limited time on the wing. In spite of this, dragonflies are often seen long distances from water.

When dragonflies and damselflies emerge from the nymph stage, their bodies are soft and the colors of the mature adult have not fully developed. They are also sexually immature, and are not ready for courtship and mating. The immature males cannot risk staying near the water, as they would be battered and damaged by adult male dragonflies, which aggressively defend their territories against other males. In addition, adult males approach any female in their territories with a view to courtship. The young dragonflies take from two or three days to three weeks to develop. Until then, they avoid the water.

Females only come to the water to mate and lay their eggs. They do not lay all their eggs at once. After laying the first batch, they avoid the attentions of the mature males by leaving the water until the next batch is ready.

LEFT The blue damselfly is a widespread insect of ponds and marshes in Europe, Asia and North America. When the larva is fully mature, it climbs a plant stem at the water's edge and its skin splits to reveal the adult insect (seen here after emerging from its transparent skin casing). The newly emerged adult is dull-colored and sexually immature.

RIGHT Reedbeds are a typical feature of swamps, and the common reed (*Phragmites communis*) is one of the most widespread plants in the world. Many invertebrates feed on reeds; some eat the foliage and others bore into the stems. Reedbeds also provide shelter and breeding sites for birds.

of barren, windswept bog, often dominated by bog-moss. Similarly, mangrove swamps and salt-marshes thrive in sea estuaries and the lower parts of river deltas.

The terms "swamp" and "marsh" both refer to types of freshwater wetlands, although there are many different names for these habitats throughout the world. They are sometimes used interchangeably and exact definitions are difficult, since there are so many variations in terrain. In the USA, swamps are usually dominated by trees or shrubs, while in Europe, marshes are characterized by tall grasses such as reeds. The two habitats may occur alongside each other.

Wetland ecology

Swamps and marshes are not permanent, unchanging places. On the contrary, they change all the time, although they do so slowly and imperceptibly. The vegetation alters according to the level of the water, and if the wetlands are fed by rivers, for example, they may dry up as the river runs low, especially in the dry season. Swamps and marshes may also be drained: this can occur directly when they are reclaimed for cultivation, or indirectly when they lose water as a result of drainage in the surrounding areas or with the extraction of groundwater.

As they grow and spread, the plants of the water's edge gradually encroach upon the wetland habitat. Under certain conditions they can gradually dry it out. At the edge of a lake, for example, plants colonize the shallow, open water. Dead organic material and silt accumulate around the bases of the plants, which slowly spread out into the open water. Their progress is assisted by the gradual build-up of sediment brought to the lake by rivers and streams.

Behind the shallow water plants, where enough material will have accumulated to dry out the ground a little, other plants establish themselves: first herbs, then shrubs and trees. The first trees to take root are water-tolerant species, but as leaf litter gathers and the ground level slowly rises, the trees that replace them are more typical of drier places. Eventually a wood forms, which surrounds and closes in on the lake and its wetland margins. The wetland, meanwhile, continues to encroach on the open water of the lake. Eventually, if these processes are allowed to

ABOVE Where wetland vegetation is short, for example in grazed marshes and wet meadows, lowgrowing plants such as marsh marigolds occur. Also known as the kingcup, the marsh marigold is a member of the buttercup family. Its large yellow flowers attract many insects, which take pollen and nectar, while helping to pollinate the flowers.

ABOVE The grass snake *Natrix natrix*, also called the water snake, is usually found in damp places. Measuring 2–3 ft long, it is an excellent swimmer and feeds mainly on amphibians, such as frogs and toads, although it takes tadpoles and fishes as well.

continue, the lake will disappear – to be followed by the wetland itself.

Not all wetlands are doomed to extinction. In many places, the local conditions prevent them drying out. If a wetland is fed by a river, for example, the flow of water may help prevent organic material and silt from building up, and the swamp is likely to remain.

A varied wildlife

The varied and changing conditions within swamps and marshes present a variety of habitats in which animals can shelter, feed and breed. Here, herbivorous animals feed on the leaves of grasses, herbs, shrubs and trees, and use the dense vegetation as cover. Carnivores of various sizes hunt the herbivores, while small birds, such as warblers, roost, feed and breed among the tall plants that stand clear of the water. Waterfowl live below them, building their nests at the bases of the plants.

Wetland invertebrates

A wide range of invertebrates live in swamps and marshland. Members of the same species also live in lakes, where the water is still and full of weed. Pond snails, pond skaters, water beetles and dragonflies are all typical of wetlands.

Some wetland invertebrates feed on the tall plants that rise above the water. In Europe, the drinker month is a typical reed-bed species, its caterpillars eating the leaves of the reeds and other grasses. The larvae of other months, as well as those of bees and other insects, feed inside the reed stems, protected from predators by their hard coverings.

ABOVE **The little egret inhabits swamps, marshes and lagoons, and feeds on fishes and aquatic invertebrates. In some parts of its range it is only a summer visitor, migrating long distances to its wintering grounds at the end of the breeding season. It was once hunted for the beautiful long white plumes of its breeding plumage.**

The myth of malaria

The insect most closely associated with freshwater wetlands is the mosquito. A biting insect, with a whining buzz, it can cause much discomfort with its bites, while certain tropical species transmit malaria. It was once believed that the mists that rose off marshes were responsible for the disease – hence the name "malaria" from the Italian *mal'aria*, meaning "bad air." Only in the 19th century was it proved that mosquitoes of some species could transmit malaria. It is because of the mosquito that many wetlands have been drained.

Wetlands, with their sun-warmed, shallow water and rich nutrient supply, are ideal breeding places for mosquitoes. The insects thrive, despite being eaten in large numbers by fish, water boatmen and beetles.

Mosquito control

Many methods have been used to reduce mosquito numbers in swamps – from spraying oil on the water, which prevents the larvae breaking the surface to breathe, to large scale spraying with insecticides. One approach is the use of biological control, using the mosquito fish.

The mosquito fish measures two inches long and is native to northern Mexico and the southern and eastern USA. It has a reputation for consuming vast amounts of mosquito larvae, and can eat its own weight of larvae in a day. On this basis, it has been introduced to mosquito-ridden areas all over the world.

Unfortunately, the mosquito fish also eats indigenous fishes that would normally eat mosquito larvae themselves, so that its arrival in new locations may be a mixed blessing. Like some other wetland fishes that live in hot regions, where water often lacks sufficient oxygen, the mosquito fish's mouth is directed upwards for taking in water just under the surface where it is richer in oxygen.

An amphibian's paradise

The world of the wetlands, being part-water, part-land, is ideal for amphibians such as frogs, toads and salamanders. In many species, the larvae develop in water, but spend their adult lives on dry land (although often in damp places).

Many amphibians migrate twice a year. In spring, large numbers of them move to the water to breed, and in the fall they migrate to suitable places to hibernate. Some species hibernate in the mud at the bottom of pools where they may also breed. Other species hibernate among vegetation or under logs or stones.

Amphibians are not restricted to swamps and marshes. In Europe, great crested newts and edible frogs breed in lakes. However, they do not occur in lakes with a large fish population, since many species of fish prey on the tadpoles. Salamanders often breed in fast-flowing streams. One of the largest of all salamanders, the Japanese giant salamander, lives permanently in rivers.

Wetland reptiles

Reptiles occur widely in marshes and swamps. In Europe, the pond terrapin is typical of well-vegetated, swampy places, where it hunts invertebrates and amphibians; it eats large numbers of tadpoles, and occasionally takes birds' eggs from nests in the low vegetation.

The grass snake and its relatives, the viperine and dice snakes, occur in European wetlands. The grass snake is the least aquatic of the three, and can sometimes be found in quite dry habitats, such as hay fields. It mainly feeds on frogs and toads. The viperine snake also feeds on amphibians, but is more tied to water than the grass snake. The dice snake is the most aquatic; it feeds on fishes, and spends most of its time hunting in the water.

Animal adaptations

Wetlands differ according to the climate and other local conditions. Their inhabitants are, therefore, forced to adapt to a variety of challenges. In temperate regions, swamps and marshes may become drier during the summer months, while in hot parts of the world, wetland habitats may disappear altogether during the dry season. When this happens, birds fly away to seek water elsewhere, while fishes swim off to find refuge in deeper waters.

The vagrant emperor dragonfly of southern Asia, Africa and southern Europe migrates vast distances when drought arrives. Dragonflies often leave the water in which they grew up as larvae and roam for a few days until they are sexually mature. Many then return to their original homes, while others colonize new sites. Those that return to the site of their original home are quite likely to find that it has dried up, since the adults breed in temporary pools that disappear very quickly in hot weather.

Wetland variety

Freshwater wetlands, including those of the USA, vary greatly in structure and appearance. Many wetland habitats occur near rivers, where they consist of vegetation interspersed with pools of water; others are simply areas of deep, water-filled hollows. Tall, swaying reedbeds, and mixed swamps with reeds, reedmace, flag iris and sedge, grow between the pools. Other wetland habitats consist of grazed marshes with low, tussock-like plants chewed down by cattle or horses. Rushes and sedges usually dominate such areas. Some wetlands embrace areas of taller marsh, with willowherbs, nettles, meadowsweet and bramble; willows, sallows and alders typically dominate wet woodland habitats.

Similar types of wetland habitat occur in different regions of the world, although the wildlife varies greatly between regions. Some sites consist of only one type of swamp and marsh, while many others embrace more than one type of habitat.

Birdlife in the wetlands

Europe's wetland habitats, such as Norfolk in Britain, are home to many species of nesting birds – the country's swamps and marshes cover a wide range of possible wetland habitat types.

Willow warblers favor wet woodland as a nesting site, since insect food abounds in the habitat's lush foliage. Other wet woodland nesting birds include deciduous woodland species such as blackbirds, bullfinches, robins, blue tits, great tits and wrens.

Snipes (wading birds with long, straight bills) are indigenous to grazed marshes – they share their habitat with curlews, lapwings, redshanks, pheasants and skylarks. Tall marshland, reedbeds and mixed reedswamp provide good nesting sites for birds such as mallards, moorhens, reed buntings, cuckoos, bearded tits, water rails and several species of warbler, including sedge warblers and reed warblers.

Different bird species use their wetland home in slightly different ways; consequently, many species are able to occupy the same habitat, since few birds compete with others for nesting space or food.

WITHDRAWN